GAME PRACTICE: CONTRIBUTIONS FROM APPLIED GAME THEORY

THEORY AND DECISION LIBRARY

General Editors: W. Leinfellner (*Vienna*) and G. Eberlein (*Munich*)

Series A: Philosophy and Methodology of the Social Sciences

Series B: Mathematical and Statistical Methods

Series C: Game Theory, Mathematical Programming and Operations Research

SERIES C: GAME THEORY, MATHEMATICAL PROGRAMMING AND OPERATIONS RESEARCH

VOLUME 23

Scope: Particular attention is paid in this series to game theory and operations research, their formal aspects and their applications to economic, political and social sciences as well as to socio-biology. It will encourage high standards in the application of game-theoretical methods to individual and social decision making.

The titles published in this series are listed at the end of this volume.

GAME PRACTICE: CONTRIBUTIONS FROM APPLIED GAME THEORY

edited by

FIORAVANTE PATRONE
University of Genoa, Italy

IGNACIO GARCÍA-JURADO
University Santiago de Compostella, Spain

STEF TIJS
Tilburg University, The Netherlands

KLUWER ACADEMIC PUBLISHERS
BOSTON / DORDRECHT / LONDON

Distributors for North, Central and South America:
Kluwer Academic Publishers
101 Philip Drive
Assinippi Park
Norwell, Massachusetts 02061 USA
Telephone (781) 871-6600
Fax (781) 871-6528
E-Mail <kluwer@wkap.com>

Distributors for all other countries:
Kluwer Academic Publishers Group
Distribution Centre
Post Office Box 322
3300 AH Dordrecht, THE NETHERLANDS
Telephone 31 78 6392 392
Fax 31 78 6546 474
E-Mail <orderdept@wkap.nl>

 Electronic Services <http://www.wkap.nl>

Library of Congress Cataloging-in-Publication Data
Game practice: contributions from applied game theory / edited by Fioravante Patrone,
Ignacio García-Jurado, Stef Tijs.
    p.cm -- (Theory and decision library. Series C, Game theory, mathematical
  programming, and operations research; v.23)
  Includes bibliographical references and index.
  ISBN 0-7923-8661-2 (acid free paper)
    1. Game theory--Congresses. I. Patrone, Fioravante.II. García-Jurado, I. (Ignacio) III.
Tijs, Stef.IV. Series.

QA269.G34 2000
519.3--dc21   99-046683

Printed on acid-free paper.

Printed in the United States of America

TABLE OF CONTENTS

PREFACE

This collection of papers is an outgrowth of the "Game Practice I" conference held in Genoa from 28th to 30th June 1998. More precisely, it is the result of the call for papers that was issued in association with that conference: actually, nearly half of the contributions to this book are papers that were presented in Genoa.

The name chosen for the conference and for this book is in evident and provocative contrast with "Game Theory": this choice needs some explanation, and to that we shall devote a few words of this Preface.

Let us say at the outset that "Game Practice" would not exist without Game Theory. As one can see, the overall content of this book is firmly rooted in the existing Game Theory. It could be hardly otherwise, given the success and influence of Game Theory (just think of the basic issues in Economic Theory), and the tremendous development that has taken place within Game Theory. This success, however, makes even more evident the existence of problems with respect to the verification of the theory. This is patent from the point of view of the predictive value of Game Theory (the "positive" side): a lot of experimental and observational evidence demonstrates that there is a large gap between theory and "practice". From the "normative" side, even if the use of Game Theory for help in advising, or in mechanism design, has seen notable successes, there is still a lot of new ground that can and should be explored by Game Theory.

There is no doubt that theory is nonetheless needed, and that it has to be developed. At the same time, there is a risk of sterile developments if theory is not fed on challenges coming from confronting the real world. In this respect, our feeling is that there is now a need for a better balance between theory and applications. The kind of considerations sketched above gave birth to the idea of organizing a series of meetings to stimulate research in Game Theory which seeks a more direct connection with real world problems.

We are well aware of the fact that defining what is a "real world application" is always a formidable task involving a careful epistemological

viii

discussion: we hope that the few sentences found in this Preface do not induce the reader to assume that the editors of this book neglect issues related with the foundations of science. Certainly we do not claim to be authorities on epistemological issues, but nor do we overlook how much theory there is between us and the "real world". Neither do we think that in the context of human behaviour, which is the key application of Game Theory, one should mimic the approach of natural (hard) sciences: it is enough to think of the problems behind the definitions of what constitutes "rational" or "intelligent" beings and behaviour.

From what we have just said, it should be clear that we do not have a formal definition of "Game Practice" to offer. On the contrary, it is perhaps worth mentioning that our opinions are significantly different. So different indeed that not all of the papers appearing in this book fit with the idea of "Game Practice" that each one of us has. Moreover, we want to add that the interesting discussion that took place at the end of the Genoa meeting neatly showed quite different points of view concerning applications of Game Theory. Anyway, an important point remains: we believe and hope that our commitment to organizing a series of meetings can have an additional non-negligible impact, together with other past and future initiatives, to stimulate research aimed at a direct and concrete verification of the theory. We have also asked Michael Maschler and Alvin Roth each to give a contribution reflecting on applied aspects of Game Theory in which they indicate briefly issues that they consider relevant for the applied side. We take the occasion to thank them for having kindly accepted our invitation, and we refer the reader to their papers, but it is worth mentioning here where they chose to put the emphasis. Maschler touches upon some issues in applying Game Theory that arise from his experience, stressing the difficulties that lie behind the choice of the right model (there is no obvious "recipe" available), together with the importance of the two-way interaction between theory and real world problems. The paper by Roth focuses on a special issue: the use of Game Theory for market design, a problem on which he has given outstanding contributions. From his experience he offers suggestions for further theoretical development, like the need for studying complex environments (so that non trivial computational work can also be important).

Looking at the contributed papers as a whole, a major problem clearly emerges, one that cannot be ignored. It is that, whenever one is interested into the applied side of a theory, it often happens that there is a significant distance between different applications. This can be seen both as a weakness (only a part of a meeting/book is of interest for each participant/reader), and also as a strength (providing inspiration, suggesting analogies or new roads, showing the richness of Game Theory). It is undoubtedly important to have more specialized meetings: this was remarked in the final discussion that ended the Genoa meeting, and various interesting suggestions were given. This was not, however, the aim of "Game Practice I", and this fact is clearly reflected by the contents of this book: just looking at the titles, one immediately realizes that the contributions address different kinds of problems.

Most of the papers can be roughly classified into four groups: political applications, problems of cost/reward sharing, economic applications and experiments. We shall now give a very concise description of the papers, beginning with the two which do not fit into this classification. The paper by Borm and van der Genutgen uses Game Theory to make a distinction, which is relevant from the point of view of law, between games of pure chance and games in which the ability of the players is relevant. In the paper by Land and Gefeller, they consider an application of the Shapley value to statistical problems in the context of medicine and insurance: reasons are given to suggest the use of a "multiplicative" variant of the value.

A couple of papers address political science issues, but from very different points of view: while Vannucci looks for institutional design problems, Brams and Togman consider the conflict in Northern Ireland, and discuss the problem of finding a model that captures the behaviour of the parties in this conflict.

Fair division problems and related algorithms are considered in the papers by Raith and by Zeng: the first focuses on multi-issue negotiation and fair-negotiation procedures, while the second focuses on approximate envy-free procedures for fair division (of cakes and chores). Fragnelli *et al.* address the issue of attributing the costs for the use of railway infrastructure, and cost sharing problems are tackled also in Feltkamp *et al.*, that consider the allocation of costs for minimum cost spanning trees, providing also a link between Bird's tree allocation and the equilibrium for an appropriate strategic form game. Equilibria (actually, pure strategy subgame perfect equilibria) are shown to exist for the so-called sequential production problems, using a potential game approach, by Voorneveld *et al.*

In the paper by Grillo the difficulties of the practical detection of collusion are discussed, while Petit and Tolwinski build dynamic models to deal with technological innovation. Auctions, and specifically the structural estimation of auction models, are the centre of interest in the paper by Hong and Shum, while van Damme makes an analysis of the 1998 Dutch auction for licences for mobile telecommunication networks. Auctions are discussed also by Parisio who deals with experiments upon bidders that do not necessarily behave as expected utility maximizers. Experimental work is also reported in the paper by Gneezy and Stoler who look for the possibility of discriminating between "revenge" and "educational" motives

x

for punishment in games like the ultimatum game.

We conclude with our thanks to the Interuniversity Centre for Game Theory and Applications, and to the Mathematical Committee of the Italian National Research Council for their financial support. Last but not least, our thanks to all of the referees for their effort, and special thanks to Renza Morino, Giulio Ferrari, Vito Fragnelli and Lucia Pusillo for their invaluable help, both for the organization of the meeting and for this book.

the editors:

Fioravante Patrone
Ignacio García-Jurado
Stef Tijs

CHAPTER 1

SOME TIPS CONCERNING APPLICATION OF GAME
THEORY TO REAL PROBLEMS

MICHAEL MASCHLER (maschler@vms.huji.ac.il)
Department of Mathematics and
The Center for Rationality and Interactive Decision Theory
The Hebrew University of Jerusalem
Israel

Let me start by saying that this note is not intended to be a survey, nor will I consider descriptive applications of game theory. The reader can find a lot of information about these topics in Aumann (1985) and in Aumann and Hart (1992 -). I shall concentrate on normal applications of game theory. More precisely, I shall deal with the issue of: how game theory can help people in pursuing their goals. I shall mainly deal with the pitfalls and how to guard against them I experienced in my brief encounters with such issues.

1. Choosing the right model

My first encounter with applications was in 1963, when I was recruited to participate in a project concerning a test ban treaty between the United States and the Soviet Union, sponsored by the U.S. Atomic Control and Disarmament Agency. The issue was how to guard against underground violations which could be confused with seismic events. Say there are n suspicious events and only r on-site inspections permitted, $r < n$. How best to allocate the permitted inspections? Basically, there are three main outcomes: "The other side violates and the violation is not caught" – utility vector normalized to (0,1) to the parties, "The other side violates and the violation is detected" – utility vector normalized to (1,0) to the parties and "no violation takes place" – utility vector equals (α, β) to the parties.

To my great surprise, all models constructed at the time ignored the possibility that no violation occurs, turning each model into a constant sum game. To me that looked absurd; the purpose of a test ban treaty is to deter the other side from conducting tests, not to catch the other side violating the treaty. By choosing a constant sum model we are pouring out the baby with the bath water. When I questioned the model I was told that the only solution game theory can offer to non-constant-sum games is equilibrium points, and we are not going to recommend equilibrium points to the U.S. Government on such an important issue. Perhaps today, when equilibrium points are more respected the answer would have been different.

I was lucky: I examined non-constant-sum variants and found out that in these cases strategies could be recommended that, up to an ϵ, are as convincing as maxmin strategies in constant-sum games. (See Maschler (1966), (1967)).

Real life is always too complicated to fully model in mathematical terms. One has to make simplifying assumptions. The moral from the above experience is that first, and above all, one should choose only such assumptions which do not destroy the important issues involved. If our simplified assumptions blur the issues then we are be in a position of the boy who is looking for a lost item under a street lamp because the place is lit, even though he knows that he lost the item in a dark alley.

If the issue is fundamentally non-constant-sum do not choose a constant-sum model. If the issue involves incomplete information among the parties, do not choose a complete information model. If the issue is basically a cooperative one (namely, binding agreements can be enforced,) be careful in your choice of the coalition-function.

2. Non-cooperative models

Eventually the test ban treaty was signed in a way that allowed underground tests, so my contribution did not help the parties. However, if they wanted to abolish underground tests, they still could not use my solution without further research. The reason is that the numbers α and β could only be estimated, at best. Some people say that because von Neumann–Morgenstern utilities cannot be measured, non-cooperative game theory involving more than 2 outcomes can only provide *insight*, but cannot actually make recommendations. I do not support this view. First, often I cannot pinpoint and say what "insight" exactly means. Second, we should strive to provide specific recommendations. I admit that use of non-cooperative game theory in the normal sense sometimes requires some sensitivity analysis and I believe that often this analysis can be supplied.

3. Equilibrium selection

What equilibrium outcome shall we recommend in each particular case? One which is Pareto undominated but uses threats that are not credible? One that is trembling hand perfect, that entails less profits to the parties? Other refinements? To provide a recommendation, we must look carefully at the real issue involved. For example, just recently Danny Granot and I discovered a perfect equilibrium profile in a certain voting scheme that is also Pareto undominated but nevertheless we are convinced that it will not be adopted by the parties. Thus, game theory offers options, but the right choice should be selected on the basis of a careful look at the real problem.

4. Cooperative games

On the one hand, many applications of cooperative games do not require measurement of utilities. On the other hand, cooperative game theory offers many solution concepts. Which one should be recommended? Some people say that we cannot recommend any particular solution, but each one provides some *insight* on the real issues. Again, often I do not know precisely what "insight" means. I also believe that we should strive to make a decision about the choice of the solution. Admittedly, this is not an easy enterprise. Let me describe a particular case where I feel that I know how to connect game theory with reality.

The issue is a bankruptcy problem $(E; d_1, d_2, \ldots, d_n)$, where an estate E is to be divided among n agents, each agent i has a claim d_i on the estate and $\sum_{i=1}^{n} d_i > E$. Present day game theory mainly recommends either the Shapley value or the nucleolus. There is also the proportional share. Which one to adopt, if any?

The nucleolus and the Shapely value can be characterized by the same set of axioms, one of which is the *consistency axiom*, also called *the reduced game property*. The only difference is in the definition of the reduced game. It is the Davis–Maschler (1965) version for the case of the nucleolus and the Hart–Mas Colell (1989) version in the case of the Shapely value. If one examines the characteristic function

$$v(S) + \max\{(E - \sum_{i \in S^c} d_i), 0\}, \quad S^c = N \setminus S, \ N = 1, 2, \ldots n, \qquad (1)$$

as was defined in Aumann–Maschler (1985), one finds that the Davis–Maschler reduced game make perfect economic sense, whereas the Hart–Mas Colell is meaningless. This was observed by Andreu Mas Colell.

I hastened to declare, therefore, that the nucleolus is the more appropriate solution for such situations. I was wrong! If we define the characteristic function by

$$
v(S) = \begin{cases} \sum_{i \in S} d_i, & \text{if } S \neq N, \\ E & \text{if } S = N, \\ 0 & \text{if } S = \emptyset, \end{cases} \tag{2}
$$

as defined in Herrero, Maschler and Villar (1999), then the Hart–Mas Colell reduced game makes economic sense and the Davis–Maschler reduced game is meaningless.

So what solution should we recommend? My present advice is to pose to the clients the following simple case: What do they think the share should be if the bankruptcy problem is, say, (238; 120, 300)? If they feel that (68, 170) is the appropriate solution, then I will recommend the proportional solution for the general problem. This is the only solution consistent with the 2-person case of this and similar examples. If they decide that (60, 178) is the right share then it seems that they regard the 300 claim as meaningless and curtail it to 238. In that case, I recommend the nucleolus of (1) as the right solution. If they choose (29, 209), that means that they regard the full 300 as the proper claim. In this case, the Shapely value, based on (2) should be recommended.[1] In all cases the consistency with the 2-person solution dictates the adopted solution.

To sum up, when we make a recommendation on a cooperative-type real issue we are faced with a problem of choosing the right solution concept. To make a good decision, we have to examine the foundations of the solution concept to see how they fit reality. The decision, however, may also depend on the coalition function we choose to model the situation. Again, we have to examine the real case in order to make a choice. It is not sufficient to recommend a certain solution on the ground that game theory has defined it. One has to justify why the particular solution is appropriate to the specific issue.

5. Theory versus real life

So far I have dealt with applications of game theory to real problems, but this is a two-way street. In 1963 I wanted to test if people really behave or at least desire to behave in accordance with the paradigm provided by the bargaining set. This experiment is reported in Maschler (1978). In

---

[1] Because of the simple form of the characteristic function it turns out that the nucleolus of (2) coincides with the Shapley value (See Herrero, Maschler and Villar (1999)). This need not be the case in more complicated bankruptcy problems; e.g., if coalitions also have claims.

that experiment I detected many cases where the bargaining set fit quite nicely, but there were also cases where the behavior was quite different. When I examined these cases I found to my surprise that, with all due respect to the bargaining set, I also would have behaved the same as some of the subjects and deviated from the bargaining set. It turned out that I embarked upon an issue which, even today, is not treated in the literature: Coalitions do not form simultaneously. Sometimes it behooves some players to rush to form coalitions. In other cases their position may improve if they wait until others form coalitions and they are left with a remainder game. This also raises the issue of transferring money in order to *encourage* others to form a coalition first and leave the game. The topic is challenging and demonstrates that real life sometimes motivates new theoretical research aimed at normative resolutions of the issues.

References

Aumann, R.J., (1985), *What is game theory trying to accomplish*, Frontiers in Economics, K.J. Arrow and S. Honkaphola, eds., Basil Blackwell, Oxford, pages 28–76.

Aumann, R.J. and S. Hart, eds. (1992 -), Handbook of Game Theory with Economic Applications, Elsevier Science Publishers (North Holland).

Aumann, R.J. and M. Maschler, (1985), *Game theoretic analysis of a bankruptcy problem from the Talmud*, Journal of Economic Theory, Vol 36, pages 195–213.

Davis, M. and M. Maschler, (1965), *The kernel of a cooperative game*, Naval Research Logistics Quarterly, Vol 12, pages 223–259.

Hart, S. and A. Mas Colell, (1989), *Potential, value and consistency*, Econometrica, Vol 57, pages 589–614.

Herrero, C., M. Maschler and A. Villar, (1999), *Individual rights and collective responsibility: the rights-egalitarian solution*, Mathematical Social Sciences, Vol 37, pages 59–77.

Maschler, M., (1966), *A price leadership method for solving the inspectors non-constant-sum game*, Naval Research Logistics Quarterly, Vol 13, pages 11–33.

Maschler, M., (1967), *The inspectors non-constant-sum game: Its dependence on a system of detectors*, Naval Research Logistics Quarterly, Vol 14, pages 275–290.

Maschler, M. (1978), *Playing an n-person game, an experiment*, Contribution to Experimental Economics, Vol 8, Coalition Forming Behavior, H. Sauermann, ed., J.C.B. Mohr (Paul Siebeck), Tübingen, pages 231–328.

CHAPTER 2

GAME THEORY AS A TOOL FOR MARKET DESIGN

ALVIN E. ROTH (aroth@hbs.edu)
Department of Economics and
Graduate School of Business Administration
Harvard University
Boston, MA 02163, USA

1. Introduction

Markets evolve, but they are also designed. Entrepreneurs and managers, legislators and regulators, lawyers and judges, all get involved, at least indirectly, in market design. Recently game theorists have also started to take a direct role in design. This is a natural development, because game theory is the part of economics that deals with the "rules of the game" that define market operations.

What makes the practical design of markets different from studying them conceptually is that practical design carries with it a responsibility for detail. For this reason, the simple models and analytical methods that characterize most game-theoretic analysis in the economic literature need to be supplemented with tools to deal with complexity. This complexity comes in two forms: complexity of the strategic environment itself, and complexity of participants' behavior.

Design practice is in its early stages, and we are learning by doing, teaching ourselves as we go. In my limited experience, some of which I will briefly describe below, a natural set of tools with which to attack complex design problems consists of analytical game theory, experimental economics, and computation. Laboratory experiments will help clarify how people behave, both when faced with new markets, and as they gain experience. Computational methods will help us explore strategic environments that may be too complex to solve analytically.

2. Market design

Market design concerns the creation of a venue for buyers and sellers, and a format for transactions. (A market as a "pure venue" can be seen in perhaps its clearest form in internet auctions, where some of the questions that arise about the location of a market are almost purely conceptual.) Game theorists have taken the lead in designing a number of different kinds of markets. Perhaps the three best known of these are auction markets for radio spectrum licenses, spot markets for electric power, and labor market clearinghouses (references: Chao and Wilson 1999, Cramton 1997, McAffee and McMillan 1996, McMillan 1994, 1995, Milgrom 1998, 1999;, Roth and Peranson 1997, 1999, Wilson, 1998, 1999). The primary design problem has been different in each of these.

In the case of radio spectrum auctions in the United States, the federal government used to give away licenses, but was ordered by Congress to sell them, both to raise revenue and to promote efficient use. The chief design question was how to design an auction procedure to promote price discovery of complicated, inter-related packages of spectrum rights in different regions. Because licenses may be more valuable in combination than separately, it was necessary to allow bidders to bid on bundles of their own choosing, in a way that allowed them to change packages in response to price changes during the auction. For this reason, multi-round auctions were adopted, and much of the design focus was on "activity rules" intended to promote efficient price discovery, by preventing bidders from concealing their interest in certain licenses and then bidding "at the last minute."

The spot market for electric power also came into being as a result of legislation, in this case to separate power transmission from power generation. Some of the chief design questions for electricity exchanges have to do with the physics of power transmission, including the negative externalities in the form of line congestion that each transaction brings to other users of the network. Thus electricity markets, in addition to mediating transactions between buyers and sellers, have to balance the transmission network geographically, and ensure the provision of adequate reserve capacity to increment and decrement the flows in response to network requirements (see Chao and Wilson, 1999).

My own experience in market design has been with entry-level professional labor markets. Since 1998, the vast majority of jobs for new physicians in the U.S. (about 20,000 per year) are filled by a clearinghouse whose design I directed (Roth and Peranson 1997, 1999). It updated and replaced an older clearinghouse, which in turn replaced a chaotic, decentralized market, which had suffered from severe coordination failures.

It turns out that these kinds of coordination failures are quite common in professional, entry-level labor markets. (And the newly designed clearing-house has in fact been adopted by entry level labor markets in a number of other professions, since its adoption by American physicians.) The demand for the design of market clearinghouses occurs in markets in which strategic behavior has led to very inefficient matching of firms and workers. The principal design task is to create a centralized market clearinghouse that will compete effectively for participants with the inefficient alternative of decentralized bilateral contracting.

To fully understand the problems that clearinghouses are intended to solve, as well as the strategic pressures that sometimes cause them to fail, it is helpful to understand the history of these various markets, and the failures they experienced that made the clearinghouse form of organization an attractive alternative. That is beyond the scope of the present paper. But some of the history of the American medical market is given in Roth (1984), some comparisons to other medical markets are found in Roth (1990, 1991), and a variety of market failures in the entry level labor markets of other professions are discussed in Roth and Xing (1994, 1997).

We can see how design problems build on and extend traditional game theoretic models by looking first at a simple model of matching, and then considering how it must be extended to deal with the practical design prob-lems involved in the medical labor market. Models of this kind have been the basis of a large theoretical literature (dating from Gale and Shapley, 1962) and a growing empirical literature. (Roth and Sotomayor 1990 survey much of this literature.)

3. A simple model of matching

There are two disjoint sets of players, Firms $= \{f_1, ..., f_n\}$, and Workers $= \{w_1, ..., w_p\}$. Associated with each firm f_i is the number of positions they have to offer, q_i. Agents on each side of the market have (transitive and strict) preferences over agents on the other side, with whom they could be matched. These are denoted by ordered lists: e.g.

$$P(f_i) = w_3, w_2, ..., w_k, \text{ for } i = 1, ..., n$$

is read as saying that firm f_i's first choice is w_3, it prefers w_3 to w_2 (denoted $w_3 \, P(f_i) w_2$), and so on, and that w_k is the least desirable worker it is willing to employ (after that it would rather leave a position vacant than fill it with a less desirable worker). Similarly, each worker w_j has preferences of the form: $P(w_j) = f_2, f_4, ...$

An outcome of the game is a *matching* x that matches firms and workers. Formally, $x : F \cup W \longrightarrow F \cup W$, such that $x(f) = w$ iff $x(w) = f$, and for all f and w, $|x(f)|$ is less than or equal to q_f, and either $x(w)$ is in F or $x(w) = w$.

A matching x will be said to be *blocked by an individual k* if k prefers being unmatched to being matched with $x(k)$, i.e. if $k\,\mathrm{P}(k)x(k)$. A matching x is *blocked by a pair of agents* (f, w) if they each prefer each other to their matches at x, i.e. if

$$w\,\mathrm{P}(f)\,w' \text{ for some } w' \text{ in } x(f) \text{ or } w\,\mathrm{P}(f)\,f \text{ if } |x(f)| < q_f$$

and

$$f\,\mathrm{P}(w)\,x(w)$$

(The first line says that f prefers w to some w' it is matched with, or f prefers w to an empty position if it has one; the second line says that w prefers f to its match at x.)

A matching x is *stable* if it isn't blocked by any individual or pair of agents. That is, a matching is stable if it assigns no agent to an unacceptable match, and if no pair of agents not matched to each other would mutually prefer to be matched to one another.

4. Some empirical evidence

Unstable matchings are hard to sustain, regardless of whether they are produced by decentralized negotiations or by centralized clearinghouses. If, for example, you are a worker holding an offer from your third choice firm, you have only to make two phone calls to find out if you are part of a blocking pair, i.e. to find out if one of the firms you would prefer to work for would prefer to employ you. As Table I shows, centralized market clearinghouses that produce stable matchings have succeeded much more reliably than those that do not. (Table I is drawn both from unpublished notes and from Roth 1984, 1991, Mongell and Roth 1991, Roth and Xing 1994, and Bodin and Pankin, 1999.)

As Table I makes clear, stability is an important feature of a centralized labor market clearinghouse. A lot more can be said about this than is evident from Table I, but I'll limit myself here to mentioning that the regional medical markets in the U.K. provide a nice natural experiment that lets us learn more about how unstable mechanisms fail, as well as how stable ones succeed. But even the best natural experiment does not permit a close examination of individual behavior (particularly when strategic behavior of various kinds may violate market rules and hence be conducted secretly). And even the best natural experiment allows for other possible interpretations-e.g. there are other differences between Edinburgh and Newcastle than the way they organized their medical markets. It is therefore

TABLE I.   Stable and Unstable (Centralized) Clearinghouses

Market	Stable?	Still in use (halted unraveling)?
American medical and other health professions labor markets		
NRMP	yes	yes(new design adopted '97)
Medical Specialties	yes	yes (about 30 markets)
Dental Residencies	yes	yes (6 markets)
Osteopaths (before '94)	no	no
Osteopaths (since '94)	yes	yes (2 markets)
Pharmacists	yes	yes
Clinical psychologists	yes (first used in '99)	yes
Regional medical markets in the U.K.		
Edinburgh ('69)	yes	yes
Cardiff	yes	yes
Cambridge	no	yes
London Hospital	no	yes
Birmingham	no	no
Edinburgh ('67)	no	no
Newcastle	no	no
Sheffield	no	no
Canadian Doctors (CaRMS)	yes	yes
Canadian Lawyers ("articling positions")		
Ontario	yes	yes
Alberta	yes	yes
British Columbia	yes	no (abandoned in '96)
Sororities	yes (at equilibrium)	yes
Reform rabbis	yes (first used in '97-98)	yes

very helpful to be able to also examine these phenomena under controlled conditions in the laboratory.

5. A laboratory experiment

Kagel and Roth (1999) report such an experiment, in which a small, decentralized laboratory market was established. It experienced costly unraveling over time in response to strategic behavior by the participants. A centralized clearinghouse was then made available. The experimental treatment was that in one condition of the experiment the clearinghouse produced stable matches (in terms of participants' stated preferences-see below), using essentially the algorithm employed in Edinburgh and Cardiff, and in the other condition, the unstable algorithm found in Birmingham and Newcastle was used. The results of the laboratory market essentially followed those observed in the field markets, and thus add an element of confidence to the interpretation of the natural experiment observed in the U.K. The control available in the laboratory, which allows us to look at markets that differ *only* in their matching mechanism, confirms that the observed changes can be accounted for by the difference between the mechanisms. The experiment also reveals some of the complexities of participants' behavior during the transition from decentralized to centralized markets that helps explain the dynamics of how centralized clearinghouses succeed or fail.

6. Stable matching mechanisms

The reason that stable mechanisms have been invented in so many markets is that stable matches can be produced by mimicking the "folk idea" of how a decentralized market operates. One version of this is stated below so as to make this clear. That is, although the algorithm described below is for a centralized clearinghouse that takes as inputs rank order lists from both firms and workers, the description is "as if" the participants were making offers and considering them, in order to make clear the analogy to a decentralized market. (Of course in a real decentralized market, we observe many varieties of strategic behavior that don't occur in the algorithm. In centralized clearinghouses, some of this strategic behavior takes place instead in the decision of what rank order lists to state. And both centralized and decentralized markets continue to promote strategic behavior at the interviewing stage. See Roth, 1991, and Roth and Xing 1994, 1997.) The algorithm described below is roughly in the form proposed by Gale and Shapley (1962).

Deferred Acceptance Algorithm, with Firms making offers:

Step 1 a. Each firm f (with q_f positions) offers them to its q_f 1st choice workers (i.e. to the top q_f workers on its rank order list).

b. Each worker rejects any unacceptable offers (i.e. offers from firms not on its rank order list) and, if more than one acceptable offer is received, "holds" the most preferred (i.e. the highest ranked).

.
.
.

Step k a. Any firm that had one or more offers rejected at step k-1 makes new offers to its most preferred acceptable workers (up to the number of rejections received).

b. Each worker holds her most preferred acceptable offer to date, and rejects the rest.

STOP: when no further offers are made, and match each worker to the firm (if any) whose offer she is holding.

It is easy to see that the resulting matching is stable, because no firm or worker is matched to an unacceptable counterpart, and if firm f, say, prefers to be matched to worker w, then firm f must have proposed to w and been rejected, so w doesn't prefer to be matched to f.

It is also easy to see that there is another, similar deferred acceptance algorithm, in which the roles of the firms and workers are reversed, i.e. in which the workers propose (or "apply" for positions) and each firm f holds its q_f most preferred applicants and rejects the rest. It can be shown that all the workers like the stable matching that results from the worker proposing algorithm at least as well as the matching that results from the firm-proposing algorithm, and that the firms similarly all do better when the firms propose.

However neither of these conclusions will hold when we extend the algorithm to deal with some of the complications found in the medical labor market. The NRMP has several kinds of complications (or "match variations"), two of which I will describe here: both cause pairs of positions to be linked as complements in the preferences of some applicants.

7. Complexities of the medical market

The first of these complications involves applicants who need two positions. Of the approximately 20,000 applicants each year, about ten percent are attempting to match to a second year position, such that, if they succeed, they will also require a prerequisite first year position. These applicants submit two kinds of rank order lists. The first is their primary list, which indicates their preferences over positions in the usual way. For each position on their primary list that requires a prerequisite first year position, they also submit a supplementary list ranking these positions.

A second, somewhat similar complication is that about five percent of the applicants are members of married couples who are seeking two residency positions in the same city. These couples submit a single rank order list, on which are ranked pairs of positions, one for each spouse. Thus a couple may indicate, for example, that its first choice is a particular pair of positions in Boston, and its second choice is another particular pair of positions in New York.

These complications matter for two related reasons:

1. They may change the properties of the market; and
2. The clearinghouse algorithm must be designed to accommodate them.

Regarding the first point, when I was asked in 1995 to redesign the clearinghouse for American physicians, I had to come to grips with the fact that the only results in the existing theory of matching that applied directly to that market were the counterexamples (see Roth and Sotomayor, 1990). The theorems were all of the form that certain things always happened in simple markets (or never happened), while the counterexamples all showed that when markets have complexities such as those described above, less regular behavior could sometimes occur. But the existing theory didn't give much guidance about magnitudes: how often different kinds of irregular behavior might occur, and just how irregular it might be. However questions about magnitudes quickly come to the fore in making design choices.

To fix ideas, below are three theorems about simple markets whose conclusions do not carry over to markets with the NRMP match variations described above.

In a simple matching market:

1. the set of stable matchings is always nonempty
2. the set of stable matchings always contains a "firm-optimal" stable matching that is produced by the firm-proposing deferred acceptance algorithm, and an "applicant optimal" stable matching that is produced by the applicant-proposing version of the algorithm.
3. When the applicant proposing algorithm is used (but not when the firm proposing algorithm is used) no applicant can possibly improve his match by submitting an ROL that is different from his true preferences.

Because these theorems about simple markets can have counterexamples in the complex medical market, we relied on computational explorations to see how close an approximation the simple theory provides for the complex market. We relied on computation in three places, each of which will be briefly described below (for details see Roth and Peranson, 1999).

8. Computational experiments to assist in the algorithm design

The process by which the applicant-proposing algorithm was designed is roughly as follows. First, a conceptual design was formulated and circulated for comment (Roth, 1996). (It was a multi-pass algorithm designed to deal with complications and resolve instabilities one at a time, and was based in part on the deferred acceptance algorithms described in Roth and Vande Vate 1990.) In order for this to be coded into a working algorithm, a number of choices had to be made, chiefly concerning the sequence in which operations would be conducted. Most of these decisions can be shown to have no effect on the outcome of simple matches, but could potentially effect the outcome when the NRMP match variations are present. Consequently, we (Elliott Peranson and I) performed computational experiments before making sequencing choices.

The results of these computational experiments on sequencing gave us our first indication that the set of stable matchings was going to turn out to be quite small. Different variations of the algorithm could indeed produce different stable matchings (in ways that would not be the case in a simple match), but it turns out that this mostly effected fewer than 2 applicants a year. That is, the differences in outcomes due to sequencing decisions in the algorithm were on the order of one in ten thousand applicants. Furthermore, the differences due to sequencing appeared to be unsystematic, and in no case did the algorithm fail to converge to a stable matching. Consequently we sequenced the algorithm in ways that cut down the number of times it cycled, and speeded convergence to a stable matching.

9. Computational explorations of the data to assess the effect of the new algorithm

Once the design of the algorithm was completed, we turned to computational explorations of the rank order lists submitted in previous years. For the moment, I will describe the analysis as if these rank order lists represent the true preferences of the medical students and residency programs, and then explain in the next section why this is not an unreasonable interpretation, for the purposes to which we put it.

In a simple market, the size of the set of stable outcomes, as measured by how many firms and workers receive different matches at different stable outcomes, can be assessed by comparing the results of the firm-proposing and the worker-proposing deferred acceptance algorithms. When this was done on the data of previous markets, we found that only about 20 applicants a year would be affected by this change, i.e. only one in a thousand.

If this were a simple market, the small number of applicants whose matching changes when we switch from hospitals proposing to applicants proposing would imply that there was also little room for strategic behavior when it comes time to state rank order lists. But this doesn't follow automatically in the complex market. And the direct computational experiment, in which each applicant's strategy choices are varied while holding all others constant, would be infeasible to conduct, given the large number of applicants in each year's data. However, for studying the strategic options available to applicants in this market, it proved possible to design efficient computational experiments that begin by truncating all applicants' rank order lists, and proceed by iteratively reducing the set of applicants whose lists are truncated, to determine upper bounds on the number of

applicants who could potentially profit from manipulating their rank order
lists. The number of applicants who could even potentially profit from such
manipulations when the applicant-proposing algorithm is employed varied
from 0 to 9 in the years examined.

10. Theoretical computation on a simple model, to study the effect of market size

The computational explorations of the data from the complex market suggest that the set of stable matchings is surprisingly small, and that the
opportunities for strategic manipulation by applicants are correspondingly
small. (The same can be said for residency programs, although I have not
discussed that here; see Roth and Peranson, 1999). But there might be
other explanations for these computational results.

   One hypothesis could be that the reason we find such small potential
for strategic manipulation is that our data have been collected after such
manipulation has already taken place. That is, perhaps there are substantial
opportunities for strategic manipulation, but these have been exhausted
by the time we look at the ROLs submitted to the match, because the
participants have already behaved strategically in an optimal way. We turn
to computational experiments to see if this is plausible.

   The computational experiments involve randomly generated simple markets with n workers and n firms, each of whom seeks to fill a single position.
The set of stable matchings will be small if the preferences are highly correlated; therefore we concentrate on uncorrelated preferences. In this case,
if agents on one side of the market have (randomly generated) preferences
over all the agents on the other side of the market, the size of the set
of stable matchings (which equals the core of the market) grows large as
the market grows large: over 90 percent of firms and workers get different
assignments at different stable matches by the time $n = 1,000$.

   However an important feature of the real market is that no one can participate in 1,000 interviews, and therefore no worker of firm submits a rank
order list over that many firms or workers. And when we fix some maximum
length k for submitted preference lists, we find that the core shrinks rapidly
as the number of firms and workers grow. Simple markets of comparable size
(in which we know we are looking at the "true", unmanipulated preferences)
exhibit the same size sets of stable matchings as we see in the NRMP, and in
smaller specialty medical markets. Consequently the size of the set of stable
matchings is a function of the market size and length of preference lists,
and the estimates on artificial markets correspond closely to the small size
observed in the data from the medical markets. This small core is a property
of the market structure (and not of manipulated preference lists). In fact

the small size of the set of stable matchings implies that there is virtually no opportunity for applicants to profitably manipulate their preference lists, regardless of what stable matching mechanism is employed.

11. Some concluding comments

If game theory is going to become as important a part of applied economics as it is a part of economic theory, we have some work ahead of us. For one thing, we'll have to develop tools to study complex environments.

Laboratory experiments will help inform us about how people behave, not only in environments too complex to analyze analytically, but also in simple environments (in which economists' customary assumptions about behavior may not always be such good approximations as we would hope). One fact that leaps out of both experimental and field data is that behavior in markets evolves, as people learn about the environment by interacting with it. And in new markets, what people learn initially is about the behavior of other inexperienced players, so that an analysis that focuses only on equilibrium behavior may miss the importance of a market's initial history. For market designers, the early history is of critical importance, since markets that fail early will not have the opportunity for players to learn about their possibly desirable equilibria.

Computational methods will help us analyze games that may be too complex to solve analytically. When game theory is used primarily as a conceptual tool, it is a great virtue to concentrate on very simple games (and even these can present formidable analytical difficulties.) When game theory is used to study empirical environments, simplicity of the models is a more nuanced virtue. But when game theory is used for designing working markets, there is no choice but to deal with the complexities that the market requires. As we have seen, computation can play different roles, from explorations of alternative design choices, to data exploration, to theoretical computation (i.e. from using computational experiments to test alternative designs, to directly exploring complex market data, to exploring related simple models in ways that nevertheless elude simple analytical solution).

In conclusion, it appears to me that the design of markets-and of other kinds of economic environments-presents both the biggest challenge and the biggest opportunity facing applied game theory at the brink of the twenty first century.

References

Bodin, Lawrence and (Rabbi) Aaron Panken (1999), *High Tech for a Higher Authority: The Placement of Graduating Rabbis From Hebrew Union College-Jewish Institute of Religion*, mimeo.

Chao, Hung-po and Robert Wilson (1999), *Design of Wholesale Electricity Markets*, in preparation.

Cramton, Peter (1997), *The FCC Auctions: An Early Assessment*, Journal of Economics and Management Strategy 6 (3), 431-497

Gale, David and Lloyd Shapley (1962), *College Admissions and the Stability of Marriage*, American Mathematical Monthly, 69, 9-15.

Kagel, John H. and A.E. Roth (1999), *The dynamics of reorganization in matching markets: A laboratory experiment motivated by a natural experiment*, forthcoming in Quarterly Journal of Economics.

McAfee, R. Preston, and John McMillan (1996), *Analyzing the Airwaves Auction*, Journal of Economic Perspectives 10, 159-176

McMillan, John (1995), *Why Auction the Spectrum?* Telecommunications Policy 19, 191-199.

McMillan, John (1994), *Selling Spectrum Rights*, Journal of Economic Perspectives 8, 145-162.

Milgrom, Paul (1998), *Game Theory and the Spectrum Auctions*, European Economic Review 42, 771-778.

Milgrom, Paul *Auction Theory for Privatization*, in preparation.

Mongell, Susan J. and Alvin E. Roth (1991), *Sorority Rush as a Two- Sided Matching Mechanism*, American Economic Review, 81, 441-464.

Roth, Alvin E. (1984), *The Evolution of the Labor Market for Medical Interns and Residents: A Case Study in Game Theory*, Journal of Political Economy, 92, 991-1016.

Roth, Alvin E. (1990), *New Physicians: A Natural Experiment in Market Organization*, Science, 250, 1524-1528.

Roth, Alvin E. (1991), *A Natural Experiment in the Organization of Entry Level Labor Markets: Regional Markets for New Physicians and Surgeons in the U.K.*, American Economic Review, 81, 415-440.

Roth, Alvin E. (1996), *Interim Report #1: Evaluation of the current NRMP algorithm, and preliminary design of an applicant-processing algorithm*, consultant's report, and http://www.economics.harvard.edu/ aroth/interim1.html

Roth, Alvin E. and Elliott Peranson (1997), *The effects of the change in the NRMP matching algorithm*, Journal of the American Medical Association, 278, 729-732.

Roth, A.E. and E. Peranson, *The Redesign of the Matching Market for American Physicians: Some Engineering Aspects of Economic Design*, American Economic Review, forthcoming.

Roth, Alvin E. and Marilda Sotomayor (1990) *Two-Sided Matching: A Study in Game-Theoretic Modeling and Analysis*, Econometric Society Monograph Series, Cambridge University Press.

Roth, Alvin E. and John H. Vande Vate (1990), *Random Paths to Stability in Two-Sided Matching*, Econometrica, 58, 1475-1480.

Roth, A.E. and X. Xing (1994), *Jumping the Gun: Imperfections and Institutions Related to the Timing of Market Transactions*, American Economic Review, 84, 992-1044.

Roth, A.E. and X. Xing (1997), *Turnaround Time and Bottlenecks in Market Clearing: Decentralized Matching in the Market for Clinical Psychologists*, Journal of Political Economy, 105, 284-329.

Wilson, Robert (1998), "Design Principles", chapter 11 in H. Chao and H. Huntington (eds.),*Design of Restructured Power Markets*, Norwell MA: Kluwer Academic Press.

Wilson, Robert (1999), "Activity Rules for an Iterative Double Auction", chapter 10 in K. Chatterjee and W. Samuelson (eds.),*Business Applications of Game Theory*. Norwell MA: Kluwer Academic Press.

CHAPTER 3

ON THE EXPLOITATION OF CASINO GAMES: HOW TO
DISTINGUISH BETWEEN GAMES OF CHANCE AND
GAMES OF SKILL? *

PETER BORM (p.e.m.borm@kub.nl)
and
BEN VAN DER GENUGTEN (ben.vdgenutgen@kub.nl)
Department of Econometrics and
CentER for Economic Research
Tilburg University, P.O. Box 90153
5000 LE Tilburg, The Netherlands

Abstract. In various countries, including the Netherlands and Austria, legislation is such
that the question whether a specific game should be considered as a game of chance or as
a game of skill is predominant in the exploitation decision of private casinos. This paper
aims for an objective and operational criterium to quantify the relative level of skill with
respect to chance of games in order to provide a juridical tool for classification. The
focus is on two-person zero-sum games. The various concepts are illustrated by means of
a variation of Poker.

1. Introduction

Does one really need skills to play the dice game backgammon or is it
basically just a game of chance? Perhaps the nice thing about this question
is the fact that almost everybody will arrive at a distinct opinion rather
quickly. For those who play backgammon frequently it will definitely be a
game of skill, in this way implicitly justifying the amount of time spent to

* This is a modified, shortened version of Borm and Van der Genugten (1998); the
main difference lies in the fact that the current paper in its quantitative analysis of skill
focusses on two-person games only. This research is sponsored by Novomatic Holland
BV, Nieuwegein. We acknowledge comments and advise from Stef Tijs and Hans Moors
of Tilburg University, of Wilfried Grossmann of the University of Vienna and of Peter
Zanoni of the Concord Card Casino in Vienna.

this game. For other people, for example for fanatic chess players, the fact that a player at every move is dependent of the throwing of dices will be inbearable and for this reason alone these people will be inclined to qualify backgammon as a game of chance. This, in our opinion, is typical for many games: opposite distinct opinions about the qualification of the game. One person will judge the level of skill of a casino game high if already one individual can beat the casino in the long run. Another person will only be convinced of the presence of skill if a vast majority of players actually displays skill. Still another person will perhaps say that by definition every casino game is a game of chance.

It is not the aim of this paper to present a general treatment of the skill concept. We restrict ourselves to a juridical point of view as reflected by gaming acts in several European countries, such as in the Dutch Gaming Act:

> .... it is not allowed to: exploit games with monetary prizes if the participants in general do not have a dominant influence on the probability to win, unless in compliance to this act, a license is granted ...

In practice the Dutch state only grants such a license to its own Holland Casino's foundation. The main reason behind this monopoly is that the exploitation of games of chance is a lucrative affair for the proprietor. The government wants to have both the control and the profits of this market.

The formulation of the Gaming Act clearly implies that skill should be considered relatively with respect to chance. Therefore we will discuss skill only in this restricted context.

In the remainder we call a game that is covered by the Gaming Act a *game of chance*. A game which is not a game of chance will be called a *game of skill*. So, in particular, to commercialize games of skill no license within the meaning of the Gaming Act is needed. This opens the way for private casinos to compete with the games of chance that are exploited in the Holland Casino's. Of course the proprietor of a game and the legislator can have different views on how to qualify a game since such a judgement about the role of chance is rather subjective and the financial stakes are high. If it would be possible to rank a broad class of games with chance elements by means of an operational and objective criterium which quantifies the level of skill relatively to chance, e.g. on a scale from zero to one, the legislator would be able to decide on a certain bound on the level of skill, below which a game should be considered as a game of chance. In this article we elaborate on a specific definition to measure skill in this way. We will focus on two-person zero-sum games. For approximations for various (one-person) casino games and a quantitative analysis for zero-sum games with three or more players we refer to Borm and Van der Genugten (1998). The underlying ideas are partly based on the analysis of Van der Genugten and Borm (1991, 1994). The various concepts will be illustrated in some detail by means of a simple variation of the game of Poker. Some general references on specific (computational) aspects on Poker are Karlin (1959a, b), Epstein (1977), Packel (1981), Scarne (1990), Sakaguchi (1993), Tamburin (1993) and Sakaguchi and Mazalov (1996). A recent paper on the topic of skill in games is Larkey, Kadane, Austin and Zamir (1997). This paper provides an interesting discussion on the interpretation and relevance of the concept of skill in analyzing and solving games. Contrary to the current exposition however the authors do not focus on a possible answer the juridical question underlying the current paper, but mainly discuss and illustrate the difference in skill incorporated in various strategies in a simplification in Stud Poker using the technique of simulation tournaments.

The paper is organized in the following way.

Section 2 discusses the qualitative requirements we think should be satisfied by a suitable relative measure of skill. The important concepts under consideration will be the learning effect and the random effect. Then, in Section 3, we discuss a definition of the level of skill for two-person zero-sum games like Chess, Schnapsen and two-person Poker. A numerical two-person variation of Poker will be used as an illustration in section 4. Section 5 concludes with the possible implications of our measure of skill.

2. Some qualitative criteria for skill

In our analysis we will restrict attention to games which, in principle, can be repeated under the same conditions. This assumption guarantees an objective quantification of uncertainty in terms of probability. This does not mean that such a quantification is impossible beforehand for games of a different kind but that it necessarily will be of a more subjective nature.

We formulate three requirements which in our opinion summarize the basic ideas underlying the Dutch legislation concerning the exploitation of games with chance elements.

R1. Legislation applies exclusively to situations which involve the exploitation of games with monetary prizes. For this reason we only consider games in which the "game-result" of a player can be expressed in some way in terms of a certain gain (or loss) of money.

R2. The skill of a player should be measured as his result in the long run, i.e. in terms of probabilistic expectation. Necessary for a game of skill is that these expected results vary among players.

R3. The fact that there is a difference between players with respect to their expected result, does not immediately imply that the underlying game is a game of skill. Sufficient for a game of skill is that the chance elements involved do not prohibit these differences to be substantial.

In Van der Genugten and Borm (1994, 1996) each of these requirements is discussed in detail in the light of the Dutch jurisprudence on specific games as Saturne, Blackjack, Roulette and Golden Ten. Implicitly these cases make clear that one should distinguish between three types of players:

(1) a *beginner* who plays the game in the naive way of somebody who has just mastered the rules of the game.
(2) a *real average player* who can be thought to represent the vast majority of players.
(3) a *virtual average player* whom we tell in advance (i.e. before he has to decide) the outcome of the chance elements.

One may conclude that legislation is concerned with the difference between a beginner and a real average player. Therefore, we call the difference in expected result between those two types of players the *learning* effect. This effect should be judged substantial in relation with the restrictive possibilities within the game set by the chance elements. This judgement we base on the so-called *random effect*: the difference in expected result between a virtual average player and a real average player. So in our view skill, as a juridical concept, reflects the relative judgement between the notions of learning effect and random effect as defined above. An objective judgement is possible if the value of the following type parameters have been determined:

- result of beginner
- result of real average player
- result of virtual average player.

The terminology above is not restricted to one-person games (e.g. Blackjack), but can be applied to two- and more-person games (e.g. Poker as well). Given the three type parameters, the following definition gives an appropriate measure for skill as a relative judgement between the learning and the random effect:

$$\text{skill} = \frac{\text{learning effect}}{\text{learning effect} + \text{random effect}}$$

$$= \frac{\text{result real average player - result beginner}}{\text{result virtual average player - result beginner}}.$$

Clearly, this implies that

$$\underset{\text{(pure chance)}}{0} \quad \leq \quad \text{skill} \quad \leq \quad \underset{\text{(pure skill)}}{1}.$$

So, games in which the random effect dominates the learning effect will have a small skill and games in which the learning effect dominates will have a high skill.

The above definition of skill makes the way in which games can be judged upon objective. It forces to think about the learning effect and the random effect in terms of the three player types above.

In order to give precise and meaningful definitions of the various concepts for two-person games we have to restrict the class of games under consideration. We briefly summarize these restrictions:

- games can be repeated

    * with objective probabilities determined by the rules of the game
    * with independency between chance elements during repetitions

- the result of a strategy of a player is based on the mean of the result in the long run (i.e. expectation) of the corresponding strategy
- the results of the different types of a player are defined under the assumption that his opponent acts as an average player
- the result of the types is the mean over the different playing roles (e.g. black and white in Chess) of a player.

Under these restrictions results of strategies of players can be described numerically. However, the type parameters reflect judgements of experts and therefore can differ. It is our experience that agreement can be obtained as far as it concerns the type beginner. However, commitment about the type average player seems to be difficult. For this we are forced to find other ways.

A very pragmatic solution is to replace the concept of the rather subjective average player with the more objective *advanced* (= almost optimal) *player*. The effect of this substitution will be an increase in the result of the real average player to the result of the real advanced player, but also an increase of the result of the virtual average player to the result of the virtual advanced player. Therefore, it may be expected that the relative change through both increases will almost remain the same. More formally,

$$\text{skill} \approx \frac{\text{result real advanced player - result beginner}}{\text{result virtual advanced player - result beginner}}$$

$$= \frac{\text{potential learning effect}}{\text{potential learning effect + potential random effect}}.$$

In the last expression we call the difference of the real advanced player with the beginner the *potential learning effect* because it does not measure what an average player will learn but what a game allows potentially to be learned. The same interpretation holds for the potential random effect as the difference between the virtual and real advanced player.

Next, the following question arises. What is the appropriate definition of the *result* of a player type? To answer this question one has to take into account not only the fact that our aim is to compare different games in a consistent way (a global argument) but also the fact that we want to have a "fair" comparison between the different types of players within a specific game with its own typical characteristics (a local argument). The underlying (local) problem mainly has to do with the fact that variation in stakes can be an important part of the strategy of a player.

Recall that we supposed (R1) that the game-result can be expressed in terms of money. The first possibility to define the result of a player is to disregard the stakes entirely and to define the result of a player-type as the (expected) *direct payoffs* arising from the game. Let P_0, P_m and P_u denote these payoffs for the three different types of players distinguished before, with the subindex 0 for beginner, m for optimal, and u for virtual optimal. A second possibility is to disregard only the variation in stakes between players in the evaluation of skill and to define the result of a player type as the expected *net gains* per instance of the game. Denoting these expected net gains by G_0, G_m and G_u and the expected stakes per instance by I_0, I_m and I_u, it is clear that $G_0 = P_0 - I_0, G_m = P_m - I_m$ and $G_u = P_u - I_u$. A third possibility would be to look at the expected *returns* which takes the variation of stakes into account, per unit of stake-money. Denoting these returns by R_0, R_m and R_u we have

$$R_0 = \frac{P_0}{I_0} = \left(\frac{G_0}{I_0} + 1\right), R_m = \frac{P_m}{I_m} = \left(\frac{G_m}{I_m} + 1\right), R_u = \frac{P_u}{I_u} = \left(\frac{G_u}{I_u} + 1\right).$$

The choice of a particular notion of result is not as innocent as it may perhaps seem: it implicitly implies an assumption on the goals of the players in a game. It will be typically the case that the strategy choice of a player (and in particular of an advanced player) will be different if he focusses on net gains or on returns. In this paper we will focus on result measured in terms of the net gains G_0, G_m and G_u. An analysis based on return is very well possible but the games under consideration become more complex. The main reason is that in a game with payoffs representing returns one looses linearity of the payoffs functions.

Before we proceed on the quantitative way in some detail, we want do discuss a more qualitative way to get some impression about skill via the learning effect and random effect by considering indicators for those effects (cf. Van der Genugten, Borm and Grossmann (1997)).

An important tool is to look at the structure of the game as it evolves in time. This will generate a game tree of all possible random moves (outcomes of chance elements) and at each stage of the game the possible moves of each player together with the history of play up to this stage and the information the player has acquired at that stage. Basically, the game tree is the structure to look at for all aspects of the game including skill. For games in practice these game trees are very complex. Roughly speaking, the more random moves the higher the random effect and the more players moves and information the higher the learning effect. These two indicators have an opposite effect on skill.

In one-person games the historical information at a certain stage of the player can only help him with the choice of his moves (e.g. remembering cards played in Blackjack). It helps him to choose a good move. We call the ability to exploit the historical information *observational skill*.

In two-and more-person games observational skill is an important ability too. However, in such games the other players not necessarily have the same historical information a particular player has when having to choose his move. A player can profit from the uncertainty of the others by varying his moves in a reasonable chosen way. This is often far more better than making each time the same move in a particular situation. We call this *strategic skill*. In fact, this way of playing can result in moves which are very bad if the other players would have the same information he has (e.g. raising in Poker with low cards). We call the ability of playing in such a way *strategic bluffing*. There is nothing mysterious about this bluffing skill: its effectiveness can be proved in a mathematical precise way. It is part of strategic skill.

There is a lot of misunderstanding about bluffing in card games, especially in Poker. Besides strategic bluffing one can observe also psychological bluffing. By this we mean misleading opponents with tools outside the formal rules of the game (e.g. playing attitude as is always strongly suggested in Western movies). We want to emphasize that in our analysis we totally disregard psychological bluffing in this sense.

Summarizing, important indicators for a learning effect are observational skill and, for more-person games, strategic skill (including strategic bluffing).

3. Two-person zero-sum games

In a two-person game typically the net gains of a player do not only depend on his own strategy choice but also on the strategy choice of the other player. In this section we will consider strictly competitive games in the sense that the gains of one player equal the losses of the other in each instance of the game. Put differently, the players exchange an amount of money specified by the outcome corresponding to one specific strategy-combination. This type of game is called a zero-sum game because the total gains of the players equal zero. For example, a Poker game as Seven

Card Stud typically can be classified as a zero-sum game: apart from an admission fee, money is reallocated among the participants.

Consider an arbitrary two-person zero-sum game where the payoffs are assumed to represent net gains in which both player 1 and player 2 can choose from a finite number of actions. In this game we will allow a player to choose a mixed strategy, which provides a probability distribution on the set of actions.

Then this game has a uniquely determined (minimax-) *value* V in the sense that player 1 can choose a mixed strategy which guarantees him an expected payoff of at least V (independent of player 2's strategy choice), and player 2 has a mixed strategy which keeps his losses down to at most V. The strategies concerned are called minimax strategies and can be interpreted as describing optimal play.

Assume we are in the role of player 1. If we fix a specific minimax strategy of player 2 in the sense that we assume that player 2 will act accordingly, the original two-person game can be reduced to an optimization problem because player 1 faces a fixed but probabilistic environment, and the three basic notions of result (w.r.t. player 1) based on net gains are well-defined: $G_0(1) \leq G_m(1) \leq G_u(1)$.

Note that, by definition, $G_m(1)$ equals the value of the game.

Analogously, in the role of player 2, we assume a specific minimax strategy of player 1 to be given, which reduces the situation to an optimization problem. This leads to the following notions of expected gains w.r.t. player 2: $G_0(2) \leq G_m(2) \leq G_u(2)$, where $G_m(2)$ equals minus the value of the game: $G_m(2) = -G_m(1)$.

To define the overall notion of skill in the game we average between the role of player 1 and player 2, implicitly assuming a player of the game to take both the role of player 1 and player 2 every two instances of the game.

This leads to

$$
\begin{aligned}
G_0 &= \tfrac{1}{2}(G_0(1) + G_0(2)), \\
G_m &= \tfrac{1}{2}(G_m(1) + G_m(2)), \\
G_u &= \tfrac{1}{2}(G_u(1) + G_u(2)),
\end{aligned}
\tag{1}
$$

and, clearly, $G_0 \leq G_m \leq G_u$.

Further note that $G_m = 0$ since $G_m(1) = -G_m(2)$. We will assume that

$$
G_0 < G_u.
\tag{2}
$$

Subsequently, we define the level of skillby.

$$S = \frac{G_m - G_0}{G_u - G_0}.$$ (3)

4. Two-person Mini Poker

Two-person Mini Poker is a game of cards played by two players, named player 1 and player 2, and with three cards of which only the numeric value is important. These values are 10, 20 and 30, respectively.

Before playing both players donate \$ 1 to the stakes. After (re)shuffling the deck of cards each player is dealt one card. Each player knows (the value of) his own card but not the card of this opponent. Thus, the one card which remains in the deck is not shown to either of the players.

Player 1 starts the play and has to decide between "checking" or "raising". If he decides to check, player 2 has to check too and "showdown" follows. If player 1 decides to raise, he has to add one extra dollar to the stakes. Subsequently, player 2 has to decide between "folding" or "calling". If the decides to fold, player 1 gets the stakes. If player 2 decides to call, he also has to add one extra dollar to the stakes and "showdown" follows. If the players have decided upon "showdown" both cards are compared and the player with the highest card value gets the stakes.

In this simple game of Poker both chance (dealing cards) and skill (w.r.t. a good betting strategy) play a role. We will reach the conclusion that the level of skill S equals 0.2.

First we will describe the possible strategies of both players. For each of the three possible cards, both player 1 and player 2 have to choose between two possible actions, leading to a total of 8 combinations. These pure strategies are represented in table 4.1 and table 4.2.

	1	2	3	4	5	6	7	8
10:	C	C	C	C	R	R	R	R
20:	C	C	R	R	C	C	R	R
30:	C	R	C	R	C	R	C	R

Table 4.1: Pure strategies of player 1. (C = Check, R = Raise)

	1	2	3	4	5	6	7	8
10:	F	F	F	F	C	C	C	C
20:	F	F	C	C	F	F	C	C
30:	F	C	F	C	F	C	F	C

Table 4.2: Pure strategies of player 2. (F = Fold, C = Call)

For example, the strategy $\underline{4}$ of player 1 should be interpreted as
  "Check with 10, Raise with 20, Raise with 30",
  and strategy $\underline{6}$ of player 2 as
  "Call with 10, Fold with 20, Call with 30",
  if you are called upon to act.

Allowing for mixed strategies, each player may put a probability measure on his set of pure strategies. Mixed strategies will be denoted in the following way: the strategy $\left(\frac{1}{2}2, \frac{1}{2}4\right)$ of player 1 will indicate that player 1 will choose the pure strategy $\underline{2}$ with probability $\frac{1}{2}$ and the pure strategy $\underline{4}$ with probability $\frac{1}{2}$. In the equivalent behavioral interpretation this boils down to:
  "Check with 10, Raise with 30, and with 20:
   Check with probability $\frac{1}{2}$ and Raise with probability $\frac{1}{2}$".

In analysing the level of skill of this game of Poker we first need the strategy of a beginner in both the role of player 1 and player 2.

W.r.t. player 1 we find it reasonable for a naive player to check with 10 and to raise with 30. How to act with 20? Probably it is wise to vary between checking and raising, and using symmetry arguments, to check and raise with equal probability. In fact, the same reasoning applies w.r.t. player 2. If he is called upon to act, we assume he folds with 10, calls with 30 and folds and calls with equal probability with 20.

Using the notation introduced above, this boils down to the beginner's strategy $\left(\frac{1}{2}2, \frac{1}{2}4\right)$ for both roles.

The next step is to determine minimax strategies for both player 1 and player 2. To this aim we first calculate the 8×8 matrix which describes the expected playoff to player 1 for each possible pure strategy combination. The result is presented in table 4.3.

pl. 2 pl. 1	1	2	3	4	5	6	7	8
1	0	0	0	0	0	0	0	0
2	0	0	1/6	1/6	1/6	1/6	1/3	1/3
3	1/3	−1/6	1/3	−1/6	1/2	0	1/2	0
4	1/3	−1/6	1/2	0	2/3	1/6	5/6	1/3
5	2/3	1/6	1/6	−1/3	2/3	1/6	1/6	−1/3
6	2/3	1/6	1/3	−1/6	5/6	1/3	1/2	0
7	1	0	1/2	−1/2	7/6	1/6	2/3	−1/3
8	1	0	2/3	−1/3	4/3	1/3	1	0

Table 4.3: Two-person Mini Poker as a matrix game

More specifically, in table 4.3 the result of a specific strategy combination is a fair average over the outcomes w.r.t. the six possible card combinations.

By iteratively deleting dominated strategies (e.g. strategy 1 of player 1 is dominated by strategy 2), the 8 × 8 matrix game of table 4.3 can be reduced to the simple 2 × 2 matrix game represented in table 4.4.

pl. 2 pl. 1	2	4
2	0	1/6
6	1/6	−1/6

Table 4.4: The reduced game

From table 4.4 it is readily derived that the value of the game is $\frac{1}{18}$ and the minimax strategies of the players are uniquely determined:

$\left(\frac{2}{3}2, \frac{1}{3}6\right)$ for player 1 and $\left(\frac{2}{3}2, \frac{1}{3}4\right)$ for player 2.

This boils down to the following (behavorial) strategy-schemes

For player 1:
   With 10: C with probability $\frac{2}{3}$, R with probability $\frac{1}{3}$
   With 20: C
   With 30: R

For player 2 (if called upon to act):
   With 10: F

With 20: F with probability $\frac{2}{3}$, C with probability $\frac{1}{3}$
With 30: C.

For determining $G_0(1), G_m(1)$ and $G_u(1)$ we assume that player 2 behaves according to his unique optimal strategy $\left(\frac{2}{3}2, \frac{1}{3}4\right)$.

Since we have chosen $\left(\frac{1}{2}2, \frac{1}{2}4\right)$ as a beginner's strategy for player 1, it follows from table 4.3 that

$$G_0(1) = \frac{1}{2}\cdot\frac{2}{3}\cdot(0) + \frac{1}{2}\cdot\frac{1}{3}\cdot\left(\frac{1}{6}\right) + \frac{1}{2}\cdot\frac{2}{3}\cdot\left(-\frac{1}{6}\right) + \frac{1}{2}\cdot\frac{1}{3}\cdot(0) = -\frac{1}{36}.$$

Further, since $G_m(1)$ equals the value of the game, we have that

$$G_m(1) = \frac{1}{18}.$$

To determine $G_u(1)$, we have to consider a virtual optimal player in the role of player 1.

By assumption, a virtual optimal player will know the precise card combination (C_1, C_2), with $C_i \in \{10, 20, 30\}, C_1 \neq C_2$, denoting the value of the card of player i, before having to decide upon checking or raising.

Moreover, since we fixed the strategy of player 2 to $\left(\frac{2}{3}2, \frac{1}{3}4\right)$ one can calculate for each card combination what the expected gains of the two possible actions will be. For each card combination a virtual optimal player will subsequently choose the action with minimal (maximal) gains. The number $G_u(1)$ then will be the fair average of the six corresponding gains. This computation, which is illustrated in table 4.5, leads to $G_u(1) = \frac{2}{9}$.

	player 2: $\left(\frac{2}{3}2, \frac{1}{3}4\right)$					
card comb.	(10, 20)	(10, 30)	(20, 10)	(20, 30)	(30, 10)	(30, 20)
Check	-1	-1	1	-1	1	1
Raise	0	-2	1	-2	1	$\frac{4}{3}$
Max	0	-1	1	-1	1	$\frac{4}{3}$
	$G_u(1) = \frac{2}{9}$					

Table 4.5: The calculation of $G_u(1)$.

In an analogous way one derives

$$G_0(2) = -\frac{1}{18}, G_m(2) = -\frac{1}{18} \text{ and } G_u(2) = \frac{1}{9}.$$

Hence,

$$G_0 = -\frac{1}{24}, G_m = 0, G_u = \frac{1}{6}$$

and, by substitution, the level of skill equals

$$S = 0.2.$$

This game of Poker is rather special in the sense that the minimax strategies of both player 1 and player 2 are uniquely determined. If this is not the case, the analysis requires a specific choice between all minimax strategies of a player. A possibility to do this would be to select what we would like to a call a *learning minimax* strategy by following a specific (modification of) learning procedure as initially proposed by Brown (1949) and Robinson (1950), and elaborated on by Shapiro (1958) and Karlin (1959a).

5. A classification of games

In this section we present a global ordering of some well-known round games w.r.t. skill. The corresponding order is partly based on a quantitative analysis along the lines sketched in this paper and partly on a (qualitative) comparison between the various possibilities of exercising skill in the underlying games. For a more elaborate discussion and motivation we refer to Borm and van der Genugten (1998) and van der Genugten, Borm and Grossmann (1997).

The main results are summarized in the following overview.

Pure games of chance

        Roulette, Craps, Trente et Quarante

        Blackjack

        Golden Ten

        Schnapsen

        Draw Poker

        Texas Hold'Em

        Seven Card Stud

        Bridge

        Chess, Checkers

Pure games of skill

The ordering between the three variants of Poker is based on the following considerations summarized in table 6.1 with respect to observational skill and strategic skill.

	Observational skill	Strategic skill
Draw Poker	-	+
Texas Hold'Em	0	+
Seven Card Stud.	+	+

Table 6.1. A qualitative judgement of skill for Poker.

Interestingly, Dutch jurisprudence has classified Golden Ten as a game of chance while according to Austrian jurisprudence Schnapsen is a game of skill (cf. Grossmann (1997)). If we combine these judgements, the decision bound between games of chance and games of skill should be laid between Golden Ten and Schnapsen. Of course, the classification given above should not be interpreted as the final order in which no shifts are possible. A further analysis, in particular w.r.t. more sophisticated variants of Poker is needed to further sharpen and quantify the classification.

References

Borm, P. and Van der Genugten, B.B. (1998). *On a relative measure of skill for games with chance elements*. To appear in Top.

Brown, G.W. (1949). *Some notes on computation of game solutions*. RAND Report P.78, The RAND Corporation, Santa Monica, California.

Epstein, R.A. (1977). *The theory of gambling and statistical logic*. Academic Press, New York.

Grossmann W. (1997). *Aspects of chance in 7-card stud* (In German). Report, Institut für Statistik, Operations Research und Computerverfahren, Universität Wien.

Karlin, S. (1959a). *Matrix games, Programming and Mathematical Economics*, vol. I, Pergamon Press, London.

Karlin, S. (1959b). *Mathematical Methods and Theory in Games, Programming, and Economics*, vol. II, chapter 9, Pergamon Press, London.

Larkey, P., Kadane, J.B., Austin, R. and Zamir, S. (1997). *Skill in games*. Management Science, 43, 596-609.

Packel, E. (1981). *The mathematics of games and gambling*. The Mathematical Association of America.

Robinson, J. (1950). *An iterative method of solving a game*. Annals of Mathematics, 54, 296-301.

Sakaguchi, M. (1993). *Information structures and perfect information in some two-person poker*. Math. Japonica, 38, 743-755.

Sakaguchi, M. and Mazalov, V.V. (1996). *Two-person HI-LO Poker- Stud and Draw, I-*
Math. Japonica, 44, 39-53.

Scarne, J. (1990). *Guide to modern Poker.* Simon and Schuster, New York.

Shapiro, H.N. (1958). *Note on a computation method in the theory of games.* Communications on Pure and Applied Mathematics , 11, 587-593.

Tamburin, H. (1993). *References guide to casino gambling.* Research Service Ltd. Mobile.

Van der Genugten, B.B. (1993). *Blackjack in Holland Casino's: hoe de dealer te verslaan!* Tilburg University Press, Tilburg.

Van der Genugten, B.B. (1997) *Blackjack in Holland Casino's: basic, optimal and winning strategies.* Statistica Neerlandica (forthcoming).

Van der Genugten, B.B. and Borm, P. (1991). *How to distinguish between games of chance and games of skill: an application to Golden Ten* (In Dutch). Report, Department of Econometrics, Tilburg University.

Van der Genugten, B.B. and Borm, P. (1994). *A comparison between games with chance elements w.r.t. skill* (In Dutch). Report, Department of Econometrics, Tilburg University.

Van der Genugten, B.B. and Borm, P. (1996). *On the level of skill and the diagnostic criteria* (In Ducth). Report, Department of Econometrics, Tilburg University.

Van der Genugten, B.B. (1997). *A method to inspect slotmachines* (In Dutch). Report, EIT, Tilburg University.

Van der Genugten, B.B., Borm, P. and Grossmann, W. (1997). *Addendum to two reports on games of skill.* Report, Department of Econometrics, Tilburg University.

CHAPTER 4

AGREEMENT THROUGH THREATS: THE NORTHERN IRELAND CASE*

STEVEN J. BRAMS (brams@is2.nyu.edu)
Department of Politics
New York University
New York, NY 10003

JEFFREY M. TOGMAN (togmanje@shu.edu)
Department of Political Science
Seton Hall University
South Orange, NJ 07079

Abstract. After briefly recounting the centuries-old dispute between Great Britain and Ireland, the current conflict over Northern Ireland between Britain and Sinn Féin/Irish Republican Army (IRA) is analyzed as a 2 x 2 game. The unique Nash equilibrium in this game is shown not to predict the recent behavior of and Sinn Féin/IRA, which declared a cease-fire in September 1994, resumed its bombing campaign in February 1996, and reinstituted a cease-fire in July 1997. However, these moves are consistent with and Sinn Féin/IRAs asserting its threat power, according to the theory of moves (TOM).

The mutually beneficial resolution of this conflict seems to lie in the farsighted strategic calculations of leaders who, recognizing that a reversion to conflict is likely to occur if conciliatory behavior is not reciprocated, reward such behavior - even if it proves costly. The effective use of threat power in the 2 x 2 game indicates a possible path toward a resolution of the Northern Ireland conflict. An appendix explores the relationship between this game and the well-known games of Chicken and Prisoners' Dilemma, based on both standard game theory and TOM.

---

* This is a revised and updated version of Brams and Togman (1998); a substantially different earlier version of this chapter appeared in Brams and Togman (1996). Steven J. Brams gratefully acknowledges the support of the C. V. Starr Center for Applied Economics at New York University. The authors thank two anonymous referees for their valuable comments.

1. Introduction

The April 1998 agreement to end the violent conflict in Northern Ireland has given many people reason to hope that a permanent peace is finally at hand. But the settlement, however encouraging, raises as many questions as it answers. How was this seemingly intractable conflict resolved? Will the compromise endure? We address these and other questions by examining the strategic situation which embroiled the Irish Republican Army (IRA) and the British government. By utilizing a dynamic approach to game theory called the "theory of moves"(Brams, 1994), we demonstrate why the IRA first suspended and then resumed its paramilitary activities, only to suspend them again, before a settlement was reached. We also draw on other major conflicts that were recently settled to indicate how farsighted leaders might prevent disruptions in their search for peace.

Some of our findings could prove helpful for those involved in future conflicts. For example, we show that threats can lead to a durable compromise, notwithstanding the common assumption that threats engender conflict rather than compromise. We argue that the use of force does not necessarily stem from a thirst for violence; in some instances it might be rational for one side to resort to force to establish the credibility of its future threats, even if that side desires peace. Our aim is not to justify such violent actions but, instead, to show how policymakers can move beyond them and parlay them into a peaceful compromise.

In this chapter we apply the theory of moves (TOM) to construct a deductive model of the strategic situation in Northern Ireland, based on a 2 x 2 game. We use this game to demonstrate why certain moves were taken by both sides, and how political leaders were able to clear a path toward peace, even though the compromise outcome in this game is not what classical game theory would consider stable. We also suggest how policymakers might utilize TOM to understand future conflicts. Finally, we indicate in the Appendix how the game-theoretic model we present is related to the well-known games of Chicken and Prisoners' Dilemma, based on both standard game theory and TOM.

2. A Centuries-Old Conflict

On February 9, 1996, after a 17-month cease-fire, the Irish Republican Army set off a bomb in East London. Less than a week later, the London police found and destroyed a bomb that the IRA had left in a telephone booth in the West End. A few days later, another IRA bomb went off on a double-decker bus. The British government, under the leadership of John Major, asserted that it would have no official contact with Sinn Féin, the political arm of the IRA, until the paramilitary activities stopped. The government also deployed 500 additional troops in Northern Ireland. In October 1996, the IRA detonated two bombs at the British Army's headquarters in Lisburn, bringing the violent conflict back to Northern Ireland for the first time since the cease-fire.

On May 1, 1997, Tony Blair became Great Britain's prime minister. His election brought a renewed sense of optimism and good will to the conflict in Northern Ireland. Responding to Blair's overtures, including his warning that "the settlement train is leaving,with or without them"(Hoge, 1997a:A7), the IRA announced on July 19, 1997, that they had "ordered the unequivocal restoration of the cease-fire"(Clarity, 1997c:1). Soon thereafter, Sinn Féin was invited to participate in the peace talks, which reconvened on September 15, 1997.

After seven months of difficult negotiations, a power-sharing accord was reached on April 10, 1998. Six weeks later, on May 23, an overwhelming 71% of Northern Ireland's voters approved the accord in a nationwide referendum. In an election held on June 26, voters gave 63% of the new Northern Ireland Assembly's seats to supporters of the accord (Hoge, 1998:4). Still, there have been sporadic incidents of violence since the referendum.

These are some of the most recent developments in the conflict over British rule in Northern Ireland. The conflict between the British government and the IRA is part of a larger struggle between Catholic Republicans, including Sinn Féin and IRA members, who want Northern Ireland to become part of an all-Ireland nation-state, and Protestant Unionists, who insist that Northern Ireland remain part of the United Kingdom.

While the British government has portrayed itself as a neutral party, prior to the election of Tony Blair it was more accurate to view it as a pro-Union force. Former Prime Minister John Major, speaking about his Conservative party in 1994, asserted, "We are a Unionist Party. We should fight for the Union"(Aughey, 1994:143). Since his election, Blair has not been as blunt about his position, but he has given no indication that he would be willing to let Northern Ireland leave the union. Indeed, Blair stated that it was unlikely that the world would "see Northern Ireland as anything but a part of the United Kingdom"(Clarity, 1997a:5).

The present conflict must be seen in the context of the centuries-old Anglo-Irish antagonism. Republicans point to the first Norman invasions of Ireland in 1169 as the start of the conflict, whereas Unionists, who favor British rule, focus on the arrival of Scottish and English settlers, beginning in 1609. The arrival of these settlers in the north, often called the "plantation of Ulster,"is the source of Northern Ireland's Protestant, and mostly Unionist, majority (57% today).

Republicans have continually fought against British rule, most fiercely at the end of the nineteenth and the beginning of the twentieth centuries. The war of independence, from 1919 to 1921, led to the Anglo-Irish Treaty of 1921, negotiated by the British government and Sinn Féin. The treaty granted independence to the 26 southern counties of Ireland, which became the Republic of Ireland, but it gave Great Britain control over Northern Ireland. Assurances the treaty offered Sinn Féin that the dispute over Northern Ireland would be resolved came to naught.

After 1921 Northern Ireland remained under British control, although the Stormont regime set up by the British did enjoy a certain degree of autonomy. Until the late 1960s, the armed Republican movement in Northern Ireland met with little support or success (O'Leary and McGarry, 1993:161). Then, in the late 1960s, Catholics in Northern Ireland began a series of civil rights marches to protest, among other things, discrimination in voting, employment, and housing. The marchers appealed to the British government to protect their rights as British citizens, but they were attacked by Unionist extremists, including some in the security forces (Rose, 1971:156). Subsequently, violence spread rapidly.

In 1969 the British government sent troops to Northern Ireland in an attempt to quell the unrest. Although Britain desired to maintain the quasi-independence of Northern Ireland, the conflict spiraled out of control; in 1972 the Stormont regime was suspended and replaced by direct rule from London. Attempts by the British government to control the violence during the 1970s and 1980s failed miserably. Since 1969 more than 3,200 people have been killed in sectarian fighting in Northern Ireland (Chepesiuk, 1997:10). Furthermore, 20,000 people have been injured, with economic costs to the British government running well over $1 billion a year (Ruane and Todd, 1996:1-2).

3. The Conflict from a Strategic Perspective

By 1994 Northern Ireland had experienced a quarter-century of widespread sectarian violence. In this section we present a game-theoretic view of the conflict between Sinn Féin/IRA, treated as one player, and the British government in the period preceding the 1998 peace agreement. There were, to be sure, other important actors in the conflict, including the Republic of Ireland, Unionists in Northern Ireland, and Republicans who were not associated with Sinn Féin/IRA. However, by focusing on the struggle between the British government and Sinn Féin/IRA, we highlight the central conflict, whose dynamics we will analyze in the next section.

We consider two basic strategic stances the two sides could take. One is a hard-line stance, denoted by H. For Great Britain, this entailed a refusal to negotiate with Sinn Féin/IRA, as well as the maintenance of British rule by force. For Sinn Féin/IRA, it meant a refusal to accept any resolution short of complete independence, taking whatever paramilitary actions were necessary to undermine British rule.

Each side, as an alternative strategy, could take a conciliatory stance, denoted by C. For Great Britain, such a stance meant a willingness to negotiate a compromise solution to the conflict, including a demilitarization of its position. For Sinn Féin/IRA, C indicated a similar willingness to compromise, including halting its paramilitary activities, at least temporarily.

Figure 1

Payoff Matrix of the Northern Ireland Conflict

Great Britain

		C	H
Sinn Féin/IRA	C	Compromise (3,3)	Capitulation by Sinn Féin/IRA (2,4)
	H	Capitulation by Great Britain (4,1)	Violent Conflict (1,2)

↑
Dominant Strategy

Key:
C = conciliatory stance
H = hard-line stance
(x,y) = (payoff to Sinn Féin/IRA, payoff to Great Britain)
4 = best; 3 = next best; 2 = next worst; 1 = worst
Outcome associated with Nash equilibrium strategies underscored

The choice of C or H by each side leads to four possible outcomes, or states, that can be summarized as follows:

1. **C - C.** Compromise, resulting in a peaceful settlement.

2. **H - H.** Violent conflict, resulting in the continuation of the "Troubles."

3. **H (Sinn Féin/IRA) - C (Great Britain).** Capitulation by Great Britain, which unilaterally withdraws its forces.

4. **C (Sinn Féin/IRA) - H (Great Britain).** Capitulation by Sinn Féin/IRA, which unilaterally stops its armed resistance.

We next rank these four states for both sides as follows: 4 = best; 3 = next best; 2 = next worst; 1 = worst. Thus, the higher the number, the greater the payoff to a player. These numbers, however, do not signify any numerical value or utility a player attaches to a state. Rather, they indicate only that each player prefers a higher-ranked state to a lower-ranked one.

In the payoff matrix shown in Figure 1, these ranks are given by the ordered pair (x,y), where x is the ranking of the row player (Sinn Féin/IRA) and y is the ranking of the column player (Great Britain). We offer the following brief justification of these rankings for each player, starting with the upper-left state and moving clockwise around the matrix: [1]

Compromise: (3,3). This is the next-best state for both players, involving a compromise on the issue of sovereignty.[2] For both Great Britain and Sinn Féin/IRA, the benefits of this state include an end to the violence and the possibility of long-term peace in Northern Ireland.

IRA Capitulates: (2,4). This is the best state for Great Britain because it has all the benefits of a compromise without having to make any concessions. It is the next-worst state for Sinn Féin/IRA because, while life in Northern Ireland achieves some level of normalcy, British rule remains in place.

Violent Conflict: (1,2). This is the next-worst state for Great Britain because, although it maintains control over Northern Ireland, paramilitary attacks continue; in addition, Britain faces pressure from the Republic of Ireland, the United States, and the European Union to bring an end to the violence. It is the worst state for Sinn Féin/IRA, because both British rule and the violence continue.

Britain Capitulates: (4,1). This is the worst state for Great Britain, which loses all control over Northern Ireland by withdrawing its forces; Britain is also seen as caving in to terrorism. By contrast, Sinn Féin/IRA achieves its best state by gaining independence without the need to compromise its hard-line position.

---

[1] We consider alternative rankings of the two worst states (1 and 2) for each player in the Appendix, showing that they lead to the symmetrical games of Chicken and Prisoners' Dilemma.

[2] Various compromises that were proposed included partitioning Northern Ireland between Great Britain and the Republic of Ireland and rule by a joint Anglo-Irish authority. To facilitate a compromise, the simultaneous surrender of weapons by the IRA and Unionist paramilitary groups was proposed by former U.S. Senator George Mitchell, who chaired the international commission overseeing peace talks from June 1996 to April 1998.

The ostensible solution to this game is the (3,3) compromise, but this is not the solution that standard game theory predicts. The reason is that Great Britain has a *dominant strategy* of H: it is a better strategy than C whatever strategy Sinn Féin/IRA chooses. If Sinn Féin/IRA chooses C, then (2,4) is better for Britain than (3,3); if Sinn Féin/IRA chooses H, (1,2) is better for Britain than (4,1).

Presuming that Britain chooses H because it is unconditionally better than C, what will Sinn Féin/IRA do? Observe that Sinn Féin/IRA does not have a dominant strategy: H is better if Britain chooses C, giving (4,1) rather than (3,3), but C is better if Britain chooses H, giving (2,4) rather than (1,2).

In a game in which all parties have complete information, we assume that Sinn Féin/IRA can anticipate that Britain will choose its dominant strategy of H. Accordingly, its best response would be to choose C, obtaining its next-worst state of (2,4) rather than its worst state of (1,2).

The strategies that yield (2,4), or capitulation by Sinn Féin/IRA, are what game theorists call a *Nash equilibrium*, because if either player departs unilaterally from its strategy associated with this state (C for Sinn Féin/IRA, H for Britain), it does worse: by changing its strategy from C to H, Sinn Féin/IRA would move the situation to (1,2), or violent conflict; by changing its strategy from H to C, Great Britain would move the situation to (3,3), or compromise. By contrast, if the players both chose C, leading to compromise, each would have an incentive to depart from C to try to achieve its best state — (2,4) for Great Britain and (4,1) for Sinn Féin/IRA.

The states of (4,1) and (1,2) are also unstable in the sense that at least one player would have an incentive unilaterally to change its strategy. Hence, (2,4) is the unique stable state in this game.

The dominance of H for Great Britain helps to explain that party's refusal to negotiate with Sinn Féin, even during the 1994-96 cease-fire. However, the actions of the IRA – commencing paramilitary activities, suspending them, resuming them, suspending them once again – belie the supposed stability of (2,4). Within the confines of classical game theory, any use of force by the IRA would seem to be irrational. (But see the alternative ranking of this game as a Prisoners' Dilemma in the Appendix, in which case the H-H state is a Nash equilibrium.)

In order to account for the changes in strategy by Sinn Féin and the IRA, we next turn to the theory of moves, which allows for strategy shifts by players as they attempt to implement desired outcomes. It also allows for the exercise of threats by a player that has the power and will to carry them out if the response it seeks from the threatened party is not forthcoming.

4. TOM and Threats

Game theory, as developed initially by von Neumann and Morgenstern, is an approach that is, in their own words, "thoroughly static" (von Neumann and Morgenstern, 1953:44). Classical game theory has little to say about the dynamic process by which players' choices unfold to produce outcomes, at least in strategic-form games that are defined by payoff matrices like that shown in Figure 1. By contrast, TOM adds a dynamic dimension by assuming that players look ahead before making a move, switching strategies in anticipation of the possible moves of an opponent.

A key concept of TOM, and one which is very helpful in analyzing the conflict in Northern Ireland, is the notion of "threat power". A player has *threat power* when it can better endure an inefficient state than can an opponent.[3] An *inefficient state* is one that is worse for both players than some other state. Thus in the Figure 1 game, (1,2) is an inefficient state, because it is worse for both players than either (2,4) or (3,3).

Consider the situation in Northern Ireland, as depicted in Figure 1, and how the two sides have attempted to assert their threat power. During most of the post-1970 conflict, the IRA used its paramilitary forces to try to establish its threat power by signaling its willingness to endure the mutually harmful (i.e., inefficient) state of (1,2). Observe that by choosing H, Sinn Féin/IRA ensures that Great Britain is faced with its two worst states, (4,1) and (1,2). Presented with this choice, Britain would presumably select (1,2) over (4,1) by choosing H as well.

By asserting its threat power, Sinn Féin/IRA took a hard-line stance — but not because it preferred the conflict at (1,2) to capitulation at (2,4). Instead, it hoped to force the British to take a conciliatory stance. As Gerry Adams, the president of Sinn Féin, put it, "The course I take involves the use of physical force; but only if I achieve the situation where my people prosper can my course of action be seen, by me, to have been justified" (Clark, 1994:79).

Recall that Great Britain has a dominant strategy of maintaining its own hard-line position (H), which is better for it whatever Sinn Féin/IRA does. But when Great Britain implements its dominant strategy at the same time that Sinn Féin/IRA threatens Britain's two worst states with its choice of H, the result is violent conflict. This state held throughout most of the 1970s and 1980s.

---

[3] A rigorous definition of threat power, which only one player is assumed to possess, and an analysis of its effects in all 2 x 2 ordinal games, is given in Brams (1994, ch. 5). Here we will illustrate its effects only in the Figure 1 game and, briefly, in Chicken and Prisoners' Dilemma in the Appendix

One way out of this situation is for both sides to agree to move to the mutually beneficial compromise state. In the early and mid-1990s there were talks to try to arrive at such a settlement. The British position was that the IRA would have to renounce its use of paramilitary activities before formal negotiations for a resolution of the Northern Ireland conflict could begin. In essence, Great Britain was insisting that Sinn Féin/IRA move from H to C first, shifting the game from (1,2) to (2,4).

On the other hand, if Great Britain moved first to C, the situation would shift from (1,2) to (4,1), at least temporarily. Then Sinn Féin/IRA could move to C, resulting in the (3,3) compromise state. But this sequence of moves could be interpreted as Great Britain's giving in to terrorism at (4,1), which was unacceptable to the British government and also entailed the risk that Sinn Féin/IRA would not subsequently move on to (3,3).[4] Hence, Britain insisted that Sinn Féin/IRA make the first conciliatory move.

Sinn Féin/IRA agreed to these conditions in September of 1994 by declaring a "total cease-fire". This can be seen as a move by Sinn Féin/IRA from (1,2) to (2,4), which is better for both players, yielding an efficient if lopsided state. In return, Sinn Féin/IRA hoped that Britain would also switch strategies to a conciliatory stance by entering negotiations with it to resolve the conflict, leading to a final settlement at (3,3).

After Sinn Féin/IRA halted its paramilitary activities, and the situation stood at (2,4), Great Britain was not responsive: it did not enter negotiations with Sinn Féin, nor did it make any significant concessions. While (2,4) is Great Britain's best outcome, from which it would have no motivation to move, Sinn Féin/IRA still possessed the threat power to move back to (1,2). In short, the threat was that if the British government did not move to a conciliatory stance, leading to (3,3), the IRA would return to a hard-line stance, reinstating the inefficient state of (1,2).

Great Britain, under the leadership of John Major, was not willing to open negotiations with Sinn Féin unless the IRA first surrendered its weapons. Thus, the situation stood at (2,4) after the IRA declared a cease-fire in 1994. The British, by demanding that the IRA go one step further and disarm itself, sought to eliminate its adversary's threat power — that is, its power to revert to H and, once again, to inflict on Britain one of its two worst outcomes. If Sinn Féin/IRA did return to H, Great Britain would continue to implement its own hard-line stance (because it preferred

---

[4] While an analysis of the foregoing sequences smacks of an extensive-form game, such a representation would not make endogenous who moves first, which the TOM threat analysis does. Moreover, even applying the extensive-form analysis in TOM to the Figure 1 game gives only (2,4) as the "nonmyopic equilibrium" (see Brams, 1994, Appendix, p. 217, game 22), which does not explain — as threat power does — the eventual choice of (3,3).

[1,2] to [4,1]), which would mean a return to violent conflict at (1,2).

By refusing to move from (2,4) to (3,3) by entering negotiations with Sinn Féin during the 1994-96 cease-fire, Great Britain may have passed up an important opportunity to achieve a lasting peace. After the cease-fire, as the expression goes, the ball was in Britain's court. Yet Britain did little. The Prime Minister of the Republic of Ireland, John Bruton, claimed that "Britain had shown less courage, generosity and decisiveness since the paramilitary cease-fires last year than had many people in Ireland"("Northern Ireland's Peace Process: The Nitty Gritty,"1995:62). To most observers, Bruton seemed to be saying that Major had not been able to reciprocate the bold action taken by Gerry Adams in declaring a cease-fire and sustaining it for nearly a year and a half.

The IRA refused to disarm, prior to any settlement, for a very good reason: disarming would deprive it of the only leverage it had and would be "tantamount to surrender"(Clarity, 1997:A4). In the absence of a Republican threat, and the resolve and wherewithal to carry it out, Great Britain had no incentive to move away from its best state of (2,4).

The international commission that former U.S. Senator George Mitchell chaired implicitly recognized this dynamic by recommending that all-party negotiations be conducted before the "decommissioning"of paramilitary arms. But Major rejected this suggestion and refused to enter ministerial-level talks with Sinn Féin.

In terms of our analysis, Great Britain was unwilling to move from (2,4) to (3,3). While Britain's stay-put strategy is rational in the short run because it enjoys its best state at (2,4), it is irrational if Sinn Féin/IRA is capable of reverting to (1,2), as proved to be the case. Indeed, Kevin Toolis argued that the IRA did not restart their bombing campaign "on a whim"(Toolis, 1996:A19). It believed, instead, that the British government betrayed promises, made in secret negotiations between 1990 and 1993, that it would be forthcoming if the IRA demonstrated good faith by renouncing violence and maintaining a cease-fire.

It was, unquestionably, Major's decision not to negotiate that persuaded Sinn Féin/IRA leaders to resume the use of violence in February 1996 (Holland, 1996). Thus, after a cease-fire which lasted nearly a year and a half, the IRA resumed its paramilitary activities by commencing a bombing campaign in London, and later by extending the violent conflict to Northern Ireland, thereby returning the situation to the destructive (1,2) state.

5. A Path Toward Peace

The IRA's restoration of its cease-fire in July 1997 offered hope that a peaceful resolution to the conflict in Northern Ireland might yet be within reach. In what follows, we analyze the moves and countermoves of Britain and Sinn Féin/IRA by applying our game-theoretic model to the 1997 efforts to attenuate the conflict — efforts that laid the groundwork for the final negotiations in 1998.[5]

Even before Tony Blair's election, it had become clear that a Sinn Féin/IRA cease-fire would have to precede any substantial conciliatory moves on the part of the British government. Like John Major before him, Blair insisted on this conciliatory move. In theory, the game we present in Figure 1 can be moved from conflict at (1,2) to compromise at (3,3) via two paths. One path to compromise would involve Great Britain's changing from a hard-line to a conciliatory stance first, followed by a reciprocal change of strategies on the part of Sinn Féin/IRA. As a result, the situation would initially move from (1,2) to (4,1), and subsequently from (4,1) to (3,3). In fact, Gerry Adams tried to see if Blair might be willing to make some concessions, even before an IRA cease-fire, by showing up uninvited at peace talks on June 3, 1997, declaring that Sinn Féin had a mandate to participate (Clarity, 1997b:A8).

But, as we have already noted, Great Britain could easily be seen as having caved in to terrorism if such a path were followed. Not surprisingly, Adams was not allowed to participate in the talks.

The second possible path toward peace would require that Sinn Féin/IRA change from a hard-line to a conciliatory stance first, followed by a recip-rocal change on the part of Great Britain. In this scenario, the situation would initially move from (1,2) to (2,4), and subsequently from (2,4) to (3,3). While the outcome would be the same, the British government under the leadership of both Major and Blair was insistent in its demand that *this* path be followed.

With its restoration of the cease-fire, Sinn Féin/IRA made the first move, shifting the situation to (2,4). This, we believe, was a necessary condition for a peaceful resolution to be reached. But, as the breaking of the cease-fire in February 1996 proved, it was not sufficient.

Sinn Féin/IRA declared a second cease-fire in July 1997 only after it became convinced that Great Britain would respond with significant con-cessions that would facilitate a lasting resolution to the conflict. In terms of our model, this means that Sinn Féin/IRA was willing to move the situation

---

[5] While the 2 x 2 game we postulate, and the rational play of it based on TOM, seem to us to illustrate how a compromise could be forged, an alternative formal analysis of this conflict is given in Miall (1996) using "drama theory". In contrast to our treatment, Miall contends that there must be *preference* changes (which drama theory allows for) on the part of a number of parties in order to achieve a peaceful settlement. As we see it, however, changes in *strategies* by the players would be sufficient to resolve the conflict, given Sinn Féin/IRA has threat power. Furthermore, strategy changes, we believe, are more empirically plausible and theoretically parsimonious than preference changes.

from (1,2) to (2,4) — temporarily giving Great Britain its best outcome — only because it believed that Great Britain would reciprocate by taking a conciliatory stance as well, thereby moving the situation from (2,4) to compromise at (3,3).

In some respects, Great Britain's unwillingness to make such concessions during the first cease-fire made declaring a second cease-fire more daunting a prospect. Although suspicious of any new British promises of accommodation, Sinn Féin/IRA, nevertherless, chose to try again.

This second attempt was largely due to the efforts of both Republican and British leaders. For his part, Gerry Adams personalized past failures, blaming John Major for the fact that Great Britain had not reciprocated Sinn Féin/IRA's last conciliatory move. Adams did this, at least in part, because he needed to convince his fellow Republicans, especially Sinn Féin and IRA leaders, that Blair could be trusted — and that the new British leadership should not be tainted by Major's past acts. In short, the message was that Blair, unlike Major, could be trusted to move from (2,4) to (3,3).

Blair also acted quickly to assure Sinn Féin/IRA that they were now dealing with a leader who had the political will — and a majority in Parliament not beholden to Northern Ireland unionists ("The IRA Cease-Fire," 1997) — to reciprocate a cease-fire and to push toward a resolution. Within weeks of his election, Blair went to Belfast to explain his position on Northern Ireland. If republican groups were willing to take a conciliatory stance, he promised, "I will not be slow in my response" (Clarity, 1997a:5).

The disarmament of the IRA was no longer a precondition to peace talks. As was true in the case of the first cease-fire, it was implausible to expect Sinn Féin/IRA to forfeit the threat of paramilitary activities until a satisfactory settlement was reached.

John Major's insistence during the 1994-96 cease-fire that the IRA give up its arms before participating in formal negotiations scuttled any hope of a compromise at that time. Blair, it seems, came to office far more cognizant of the fact that Sinn Féin/IRA would refuse to disarm before a settlement was reached; he also proved himself sensitive to Sinn Féin/IRA's view that one of the most important functions of its weapons was the defense of the Catholic population from Protestant attacks. Blair's peace initiative of June 1997 stipulated that disarmament would *not* be a precondition to talks, that disarmament talks should be held *simultaneously* with peace negotiations, and that *both* Protestant and Catholic groups would, eventually, have to surrender their arms (Hoge, 1997a:A7).

The issue of disarmament shaped up to be one of the most difficult obstacles to overcome. The Reverend Ian Paisley, head of the hard-line Democratic Unionist Party, complained that Sinn Féin was being offered a chance to join the peace talks without relinquishing "one weapon" (Hoge,

1997a:A7). Some considered Sinn Féin to have "bombed its way to the negotiating table"(Clarity, 1997a:1, 6). But for the reasons spelled out earlier, no one expected Sinn Féin/IRA or the Unionists to disarm before a final settlement. The participants committed themselves only to "consider and discuss"the issue of disarmament (Hoge, 1997c), which has yet to be taken up seriously.

Before the 1998 agreement was reached, peace in Northern Ireland seemed impossible, rather than difficult, to achieve. Much rests on what transpires in the near future. If Blair, Adams, and other leaders involved in Northern Ireland — including David Trimble, the moderate head of Northern Ireland's largest Protestant party, the Ulster Unionists — can consolidate the peace they have forged, their efforts will surely be worthy of Nobel peace prize consideration. Then it may be said, perhaps paradoxically, that the exercise of threats helped bring an end to the conflict.

6. Game Theory, TOM, and Policymaking

It is always difficult for adversaries to move from hard-line positions to conciliatory ones. When, after years of struggle, leaders of two hostile groups are able to find the will and to develop the trust to make such moves, an historic peace can be achieved. Such seems to have been the case in South Africa, where the prime minister, F. W. de Klerk, and the leader of the African National Congress (ANC), Nelson Mandela, found a path to peace in 1990-91. A similar reconciliation occurred in the Middle East in 1993, when the head of the Palestine Liberation Organization (PLO), Yasir Arafat, and Yitzhak Rabin and Shimon Peres, the prime minister and foreign minister of Israel, negotiated a settlement — though shaky today — between the PLO and the state of Israel.

The use of threats by the ANC and the PLO were critical in pushing the process toward a compromise. At the same time, the leaders of the South African and Israeli governments were farsighted enough to see that there was a way out of their unremitting struggles. Indeed, Mandela, de Klerk, Arafat, Rabin, and Peres were all awarded the Nobel Peace Prize for the courage, generosity, and decisiveness they showed in persevering against formidable odds to reach a settlement.[6]

Realistically, not every international conflict will end in a durable settlement. What game theory and TOM can do is give insights into *possible* paths to peace, as well as into difficulties that may be encountered along

---

[6] The efforts of Betty Williams and Mairead Corrigan (founders of the Northern Ireland Peace Movement that was later renamed the Community of Peace People) to bring peace to Northern Ireland were also recognized by the award of the Nobel Peace Prize in 1976.

the way.[7] Such insights can help political leaders predict both the dynamics of conflict and the conditions that can provide an escape from it.

Conceptualizing a conflict as a game elucidates the choices policymakers face and the consequences of those choices. Because the outcome of a game depends on the choices of all players, game theory and TOM highlight the fact that a player typically does not have a unilateral best choice — that is, a dominant strategy that is best in all situations.

From a conventional game-theoretic perspective, Great Britain's choice of a hard-line stance (H) in the Figure 1 game was its dominant strategy and, therefore, would be its expected choice regardless of what Sinn Féin/IRA does. TOM pushes this analysis farther by demonstrating that Sinn Féin/IRA possessed a threat. By incorporating this aspect into the analysis, TOM helps explain (i) why Sinn Féin/IRA opted for C, at least for a while, (ii) why Sinn Féin/IRA reverted to H, and (iii) why both sides eventually chose C.

To apply TOM, we suggest, as a first step, that policymakers attempt to write down the game or games they believe are being played. This may be more complicated than it sounds. Even if we limit ourselves to relatively simple two-player, two-strategy scenarios, we find that there are 57 different *conflict games* — games in which there is no mutually best (4,4) outcome. Because there are no outcomes in conflict games that completely satisfy both players, cooperation, while possible, may be difficult to achieve.

Without appropriate tools to analyze these games, it is by no means obvious what strategic possibilities are open to players. One needs a theory that shows which outcomes are stable, from which the players would have no desire to move lest they end up, immediately or eventually, worse off. One also needs to know which outcomes are vulnerable to threats. TOM enables one to identify these properties of outcomes in the 57 conflict games, and it facilitates generalizations across all these games.

At a minimum, this theory can help policymakers distinguish better from worse strategy choices, whatever game is being played. In addition, it offers answers to such questions as: Will threats work? What kind? Which outcomes are nonmyopically stable in the sense that if players are not restricted to making just one move from an outcome, what series of moves and countermoves might they make to try to do better?

Emphatically, TOM cannot save players from a destructive conflict if a cooperative outcome is vulnerable to threats or is otherwise unstable. On the other hand, because TOM enables one to determine what changes

---

[7] While this chapter explores the dynamics of the Northern Ireland conflict in hindsight, see Brams and Togman (1996), published before Blair's election, and Brams and Togman (1998), published before the accord was signed, for analyses that use TOM to predict what steps might lead to peace.

would need to be made in order to bring relief to such a situation, it may suggest a path to peace. In this way, TOM can bring both realism and hope to a situation: it offers a clear-eyed view of the nature of conflict and, at the same time, it indicates the possibilities for transforming it into a cooperative outcome.

7. Appendix: Comparison with Chicken and Prisoners' Dilemma

The 2 x 2 game in Figure 1 that we used to model the Northern Ireland conflict combines certain properties of the well-known games of Chicken and Prisoners' Dilemma. In the Figure 1 game, the row player (Sinn Féin/IRA) has the same preferences as those found in the Chicken game, while the column player (Great Britain) has Prisoners' Dilemma preferences, as can be seen from a comparison of these three games (see Figure 2).

Both Chicken and Prisoners' Dilemma are *symmetrical games*: the players rank the diagonal outcomes the same, and the off-diagonal outcomes are mirror images of each other. In both these games, the row and column players are interchangeable, because they face the same strategic choices: what is rational for one player is also rational for the other. In the Figure 1 game, by contrast, the row and column players face different strategic choices.

In Chicken, neither player has a dominant strategy and, as a result, it is impossible to predict which of the two pure-strategy Nash equilibrium outcomes, (4,2) or (2,4), will be selected (if either). In Prisoners' Dilemma, on the other hand, both the row and column players have dominant strategies (their second strategies), yielding the unique Nash equilibrium outcome of (2,2). This Nash equilibrium, however, is inefficient, because (2,2) is worse for both players than (3,3).

In the Figure 1 game, the players are in a different predicament from that posed by Chicken or Prisoners' Dilemma. Only the column player has a dominant strategy (its second strategy). According to classical game theory, the row player should anticipate this choice and should choose its own first strategy, resulting in the Nash equilibrium outcome of (2,4) that is underscored.

In none of these three games does the compromise outcome of (3,3) constitute a Nash equilibrium. In Prisoners' Dilemma, the outcome that classical game theory predicts, (2,2), is worse for both players than (3,3), and in Chicken the two predicted outcomes [(4,2) and (2,4)] lead to a best outcome for one player but a next-worst outcome for the other. The Figure 1 game gives a similar lopsided result [(2,4)], but one that favors only the column player.

Figure 2
Payoff Matrices of Three Games

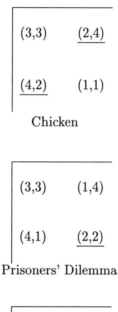

(3,3)   (2,4)

(4,2)   (1,1)

Chicken

(3,3)   (1,4)

(4,1)   (2,2)

Prisoners' Dilemma

(3,3)   (2,4)

(4,1)   (1,2)

Figure 1 Game

Key:
(x,y) = (payoff to row player, payoff to column player)
4 = best; 3 = next best; 2 = next worst; 1 = worst
Outcomes associated with Nash equilibrium strategies underscored

According to TOM, however, the compromise outcome of (3,3) in all three games can be achieved through the use of threats. Each player can threaten to choose its second strategy, associated with its opponent's two worst outcomes (1 and 2), if its opponent does not choose its first strategy when the threatener does. If this threat it credible, both players will choose their first strategies, producing the (3,3) outcome, which is clearly better for the threatened player than a 1 or 2 outcome.

As we pointed out earlier, Sinn Féin/IRA made such a threat but failed to induce (3,3) during the 1994-96 cease-fire, because Great Britain refused to heed that threat. Sinn Féin/IRA then carried out its threat and resumed paramilitary activities, resulting once again in conflict at (1,2).

The "deterrent threat"that induces (3,3) in the Figure 1 game can be undermined by a "compellent threat"(Brams, 1994, ch. 5). Specifically, the column player, by sticking with its second (dominant) strategy, can compel the row player to choose between the inefficient (1,2) and the efficient (2,4). If the column player has "threat power,"the row player can thereby induce the choice of (2,4), which is also the unique Nash equilibrium outcome. Similarly in Chicken, if the row player has threat power, it can induce (4,2), whereas if the column player has threat power, it can induce (2,4), which are the two pure-strategy Nash equilibrium outcomes in this game.

Neither player has a compellent threat in Prisoners' Dilemma, so there is nothing to undermine a deterrent threat, on the part of either player, that can induce (3,3). While it is true that compellent threats might interfere with the choice of (3,3) in both the Figure 1 game and Chicken, a deterrent threat can, in principle, work to induce (3,3) in these games. In the Northern Ireland conflict, it seems that the IRA deterrent threat, after its failure under John Major, succeeded in inducing Tony Blair to choose C, yielding the compromise (3,3) outcome.

References

Aughey, Arthur (1994). "Conservative Party Policy and Northern Ireland." In Brian Barton and Patrick J. Roche (eds), *The Northern Ireland Question: Perspectives and Policies*. Aldershot, UK: Avebury.

Bardon, Jonathan (1992). *A History of Ulster*. Belfast: Blackstaff.

Barton, Brian, and Patrick J. Roche (eds.) (1994). *The Northern Ireland Question: Perspectives and Policies*. Aldershot, UK: Avebury.

Brams, Steven J. (1994). *Theory of Moves*. Cambridge, UK: Cambridge University Press.

Brams, Steven J., and Jeffrey M. Togman (1996). "The Dynamics of the Northern Ireland Conflict." *Oxford International Review* 7, no. 2 (Spring): 50-54.

Brams, Steven J., and Jeffrey M. Togman (1998). "Cooperation Through Threats: The Northern Ireland Case." *PS: Political Science and Politics* 30, no. 1: 32-39.

Chepesiuk (1997). "The Weight of History: Ulster's Troubling Standoff." *New Leader* 80, no. 4 (March 10): 10-11.

Clarity, James F. (1997a). "Blair Makes Offer to Renew Contact With I.R.A. Wing." *New York Times* (May 17), pp. 1, 5.

Clarity, James F. (1997b). "Ulster Talks Resume With Warning That Violence Can End Them." *New York Times* (June 4), p. A8.

Clarity, James F. (1997c). "I.R.A. Announces A New Cease-Fire Beginning Today." *New York Times* (July 20), pp. 1, 9.

Clarity, James F. (1997d). "Sinn Féin Is Invited by Britain to Join Peace Talks." *New York Times* (August 30), pp. 1, 6.

Clarity, James F. (1997e). "I.R.A. Offers Support (of a Sort) for Nonviolence." *New York Times* (September 12), p. A4.

Clark, Liam (1994). "Contemporary Republican Politics." In Brian Barton and Patrick J. Roche (eds), *The Northern Ireland Question: Perspectives and Policies*. Aldershot, UK: Avebury.

Hoge, Warren (1997a). "Blair Offers New Ulster Deal; Key Is Disarming Both Sides." *New York Times* (June 26), pp. A1, A7.

Hoge, Warren (1997b). "Sinn Féin Leader Says He Expects I.R.A. Cease-Fire." *New York Times* (July 19), pp. 1, 5.

Hoge, Warren (1997c). "New Truce, New Questions." *New York Times* (July 22), p. A8.

Hoge, Warren (1998). "Vote for Assembly Realigns Northern Ireland Loyalties." *New York Times* (June 28), p. 4.

Holland, Jack (1996). "October Surprise!" *Irish Echo* (February 21-27), pp. 1, 39.

"The I.R.A. Cease-Fire" (Editorial). *New York Times* (July 21, 1997), p. A16.

Miall, Hugh (1996). "Drama Theory and the Northern Ireland Peace Process." Preprint, Department of Politics and International Relations, Lancaster University, UK (June 26).

"Northern Ireland's Peace Process: The Nitty Gritty" (1995). *The Economist* (November 18), pp. 62-63.

O'Leary, Brendan, and John McGarry (1993). *The Politics of Antagonism*. London: Athlone Press.

Rose, Richard (1971). *Governing without Consensus: An Irish Perspective*. Boston, MA: Beacon Press.

Ruane, Joseph, and Jennifer Todd (1996). *The Dynamics of the Conflict in Northern Ireland*. Cambridge, UK: Cambridge University Press.

Toolis, Kevin (1996). "Why the I.R.A. Stopped Talking." *New York Times* (February 21), p. A19.

von Neumann, John, and Oskar Morgenstern (1953). *Theory of Games and Economic Behavior*, 3rd ed. Princeton University Press.

CHAPTER 5

THE DUTCH DCS-1800 AUCTION

ERIC VAN DAMME (eric.vandamme@kub.nl)*
CentER
Tilburg University
P.O. Box 90153
5000 LE TILBURG, The Netherlands

Abstract. In February 1998 the Dutch government auctioned licences to operate mobile telecommunications networks according to the DCS-1800 technology. Two "national" licences and sixteen "regional" ones were auctioned by using a variant of the simultaneous, multiple round auction that was proposed by US-economists and that had been tested in the US. This paper describes how the decision to auction came about, it details the auction rules, and it analyzes the resulting outcomes.

1. Introduction

The increasing demand for (use of) frequency spectrum has made governments aware of the fact that spectrum is scarce and that traditional methods of spectrum allocation (on the basis of the first come first served principle, by lottery, or by means of a beauty contest) are inadequate. Worldwide it has been argued that the market mechanism could be used to ensure more efficient allocation and use of spectrum. The first spectrum auctions, in Australia and New Zealand, however, made use of traditional first price or Vickrey formats, and proved to be a mixed success. Following the advise of economic theorists and game theorists, the US-government decided to adopt a new design to auction spectrum, the multi round simultaneous auction. Using this auction format, the US-government has raised several billions of dollars. These auctions have been well documented and they have been described as a success for all parties involved. The new

---
* The author thanks Emiel Maasland and Pieter Ruys for comments on an earlier version. The paper also benefitted from discussions during conferences in Mannheim (ESA), Barcelona (Spanish Game Theory Meeting), Genoa (Game Practice Meeting) and Berlin (EEA).

format has also sparked a new wave of economic theorizing (Cramton (1995, 1997), Mc Afee and Mc Millan (1996), Mc Millan (1994, 1995), Milgrom (1996, 1998)).

In 1995, the Dutch government decided to make use of the auction mechanism to allocate spectrum rights for commercial use and, in February 1998, the first such auctions were held. In these, licences to use frequencies in the DCS-1800 band (usable for mobile telephony) were auctioned. It was decided to make use of a variant of the simultaneous multi round ascending auction that had proved so successful in the US. However, the auction rules differ in important details from those that were used on the other side of the Atlantic. In this paper, we describe the auction rules as they were used in the Netherlands and discuss the implications of the differences. We also present and analyze the actual results of the auction.

While the Netherlands is not the first European country to sell spectrum licences, it is the first European country to generate a substantial amount of money from them. (See Keuter and Nett (1997) for a description of early spectrum auctions in Germany.) Given that a good design is crucial for the auction to be a success and given the consensus that the US-auctions were a success exactly because of the fruitful interaction between academic economists and government officials, it is interesting to investigate the interaction between academics and practitioners in the Netherlands. Why did the Dutch government agency not implement exactly the same rules as in the US? What was the rationale for the changes? What role did academic economists play in the process? This paper answers these questions.

The process which led to the auction is interesting in particular because of the interaction that took place between the Dutch government and the European Commission. The latter played the leading role in the liberalization of European telecommunications markets by pushing strongly in favor of eliminating any advantages of incumbent telecommunications operators. In various directives, the Commission has argued for asymmetric regulation so as to enable a "level playing field" between incumbents and newcomers. (See especially the directive 96/2/EG.) The exact meaning of the phrase, however, is not completely clear and asymmetric treatment may violate basic principles of equal treatment. We will encounter a concrete example below. Secondly, the Commission has stated its reservations against using auctions to allocate licences, this because of the fear that the auctions might eventually result in higher prices for users. While one may doubt the economic underpinnings of this argument, it certainly played an important role in the political discussions and might have blocked the use of the auction mechanism completely. Thirdly, the Commission is suspicious of national governments creating artificial scarcity and it thus basically forces governments to make available all spectrum that is not used. As we will see, because of all these aspects, the European Commission had a crucial influence on the amount of spectrum supplied by the government, on the auction rules and, hence, on the final allocation.

The remainder of the paper is organized as follows. Section 2 describes the lengthy process leading to the Dutch DCS-1800 auction. Section 3 describes supply and demand in the Dutch DCS-1800 market and details the auction rules, stressing the important differences with the PCS-auctions as they were played in the US. Section 4 analyzes the auction outcome and some peculiarities that were observed during the play. Section 5 concludes.

2. The process leading to the auction

In this section, we briefly describe the lengthy process that eventually led to the Dutch DCS-1800 auction. We focus on the role that game theorists and economists played in this process and on the interaction between political decision making bodies at the national and European level, specifically on the intervention by the European Commission that led to changes in the auction rules.

The Netherlands was a relative latecomer in the field of digital mobile telephony. The incumbent fixed telephony operator (KPN) was given (for free) a licence in 1994, at about the same time as it was privatized. In 1995, a "beauty contest" was organized to award a second GSM-licence. The licence was won by Libertel, a join venture in which Vodaphone and the Dutch ING-bank were the main shareholders. While also this licence was awarded for free, the tender document contained statements that the winner might be charged ex post for its licence.

The selection of the winner of the beauty contest involved a lot of controversy: the selection process had taken a long time, it was not clear that the "best" party had been selected, and several parties threatened with law suits against the government. This negative experience led the government to investigate the possibility of using auctions to allocate spectrum. Informed by the positive PCS-auction experiences in the US, parliament agreed and started the process of law changing to enable such auctions. Quite naturally, Dutch economists were keen to disseminate the positive news from the US. Economists from the University of Amsterdam were the first to do so (Bykowski et al, 1995).

At that time, however, relatively little knowledge about auctions was available in the country. A good illustration is provided by a government document from 1993 in which we read the following description of the Vickrey auction: "the one who has made the second highest bid receives the lot; this prevents excessive bids being made" (HDTP, 1993, p. 39). The game that is described here is interesting, and it prevents excessive bids, but it is not the Vickrey auction. We find similar misunderstandings in later parliamentary texts. During 1996, I was asked by people from the Ministry of Finance to advise on auctions. The questions asked ranged from very general ones (pros and cons of auctions?, do auctions lead to higher consumer prices? how to prevent collusion?) to specific ones about actual auction design. I got similar questions from the Ministry of Economic Affairs and wrote a report (Van Damme, 1997a) that made the results from the literature available to a wider audience. A summary (Van Damme, 1997b) was published in the widely read weakly magazine ESB under the title "10 misunderstandings about auctions". Indeed, misunderstandings were widespread, even among the consultancy firm that was hired by the government to advise on the actual auction design. (See my letter contained in the consultation document DDV 1996.)

Meanwhile, the government had prepared a law to allow auctioning the DCS-1800 spectrum. The law stipulated that one new licence be auctioned, that incumbents be excluded from this auction and that incumbents would pay a fee, related to the auction price, for the licences they had gotten for free in the past. All these aspects, as well as the auction itself proved controversial. We briefly touch upon the reasons why, for details we refer to the parliamentary texts. (See second and first chambers of parliament, numbers 24095 and 25171.)

While parties generally accepted that auctions have attractive features, some parties nevertheless argued against auctions because they feared that they would lead to higher consumer prices and would slow down investment. These views, while incorrect according to standard economic theory (McMillan 1995), echo those of the European Commission. In particular, see the Commission's Green Paper concerning a common approach to mobile and personal communications in the EC, point 40, p. 23.

In order to avoid that governments exploit their monopoly power by creating artificial scarcity, EC Directive 94/96/EG concerning mobile communication states that the number of licences can be limited only in case of essential capacity constraints. Hence, the Dutch government could limit supply to one licence only in case no more spectrum was available. As we will see below, the EC, however, forced the Dutch government to sell considerably more spectrum that it had intended at first.

The government argued that exclusion of the incumbents was compatible with EC directive 96/2/EG concerning mobile and personal communications (Pb EG L20, 26-1-96). After all, this directive states (in consideration 8) that exclusion is allowed if otherwise there could not be effective competition, in particular because a dominant position would be strengthened. Now, at the time of the auction, KPN had a market share of about 2/3 and Libertel about 1/3, so that Libertel certainly did not have a dominant position. It is thus not surprising that Libertel was to challenge its exclusion. (See below.)

Finally, the ex post levy on incumbents was justified by the desire to create a "level playing field", newcomers would be disadvantaged if they had to pay for licences where incumbents had gotten similar licences for free. The law proposed an explicit formula linking the levy to the auction price, based on the principle that one MHz of GSM-spectrum has the same value as one MHz of DCS-1800 spectrum. Lawyers argued that the ex post levy was unjustified (that it was in conflict with the basic principle of trust) as it could not have been foreseen, while the economists argued that such a fixed payment does not contribute to create a level playing field (sunk cost are irrelevant for market behavior). For example, Sweder van Wijnbergen made this argument on behalf of Libertel.

Despite all this criticism, the responsible minister, however, saw little reason to change the law. There is only one change that she accepts: instead of one licence, two licences will be auctioned. The reason for this change are international agreements which imply that additional spectrum (in the extended GSM-band) can be allocated to mobile operators, hence, more capacity is available. With one further minor amendment, the second chamber unanimously approved the new law on June 17, 1997.

Less than 2 weeks later, however, EU-commissioner Van Miert writes a letter to the minister in which he requests not to apply the new law to the upcoming allocation of DCS-1800 licences. He proposes to use auctions only for new markets. Van Miert points out that if the new law is used to assign DCS-1800 spectrum, one might receive complaints from new or incumbent operators that might be bad for the development of the market. Indeed already on July 8, 1997, Libertel issues a complaint with the Commission. Libertel contests the scarcity levy, its exclusion from the right to bid on the DCS-1800 frequencies, and it objects to the fact that only a limited amount of spectrum is brought to the market.

On September 1, Van Miert writes the Minister that, in his first opinion, the complaint of Libertel is hard to refute: in order to be foreseeable, the levy should have been specified in a more detailed way when Libertel's licence was awarded, exclusion of Libertel cannot be justified on competition grounds, and spectrum which is available, or which will become available soon, has to be offered in the same auction. Van Miert thus urges the Minister to consider an alternative procedure. On October 7, the Minister reports to the parliament about this and she decides to change the law. In fact, the government gives in on all three points: there will be no scarcity levy (not for Libertel, neither for KPN), more spectrum will be auctioned, and incumbents will not be allowed to bid on parts of this spectrum.

For several reasons the exchange of views on the "level playing field" is interesting. At first, the Minister saw the need to levy a scarcity fee in order to comply with EU-directives. Indeed, earlier the European Commission had forced the Belgian, Italian and Spanish governments to charge their former telecom monopolists for the GSM-licences they had gotten for free after a second GSM-licence had been sold on the market. For example, Proximus (the mobile division of Belgacom) initially paid only Bfr. 3.5 million for its licence, while Mobistar (the second entrant in Belgium) had to pay Bfr. 9 billion. Van Miert insisted that Belgacom be forced to pay Bfr. 5.5 in addition. According to the Belgian newspaper De Standaard (20 Sept. '95) Van Miert said "The Commission has always stated clearly that it should be possible for newcomers to enter the market under the same conditions, hence, Belgacom knew what it could expect." Now, however, it turns out that the Commission objects to such a fee, hence, newcomers cannot enter under the same conditions. This raises the question of why the Dutch case is different? Again not surprisingly, newcomers have since complained to the Commission that the asymmetric payments violate the EU-principles and they have argued that they not be forced to pay the auction price.

All this, however, implies that the ministry gets under extreme time pressure for, after all, the EC-directive 96/2/EG stipulates that DCS-1800 licences have to be issued before January 1, 1998. The minister gets the parliament to cooperate in a quick procedure to change the law. The discussion in parliament is mainly political and does not focus on content. As far as content is concerned, discussion focuses mainly on old issues (such as do auctions lead to a higher price) and there is very little discussion about the important issue about how the supply of spectrum has to be packaged. Even though some parties argue that it would be desirable to discuss the actual auction design, one realizes that time is too short for that and MP's are satisfied after the Minister has repeated her earlier claim that an extensive process of consultation has shown that a multi round simultaneous auction is to be preferred. After the superficial discussion, both chambers of parliament accept the new (revised) law. The second chamber accepts it (unanimously) on November 18, 1997 and the first chamber, (also unanimously) on November 25, 1997. In the next subsection, we describe the final version of the law.

One may doubt, however, whether the consultation process was as extensive as the Minister claimed it to be. As far as I know, the only serious investigation about the proper design was done in the Summer of 1996 by researchers from the CREED laboratory in Amsterdam. The researcher's report (Olsen et al (1996)) is dated September 30, 1996 and is based on the assumption that only one licence will be auctioned, see p. 15, hence, it does not deal with the intracies of how to auction multiple goods with complementarities. Indeed, I was told later that, in the end, a high official at the Ministry wrote down the rules of the game on one Friday afternoon.

3. The auction rules

In this section we describe the lots that were auctioned, the auction rules and the bidders. The ministry published the detailed regulations concerning the auction on November 26, 1997, immediately after the new telecommunications law had come into effect on November 25. Interested parties had to register before mid January. The auction itself started on February 12, 1998, and ended on February 26 after 137 rounds. There was a lot of secrecy involved, only after the first round of the auction did a player find out whether all reportedly interested parties were indeed eligible to bid, and on which lots.

3.1. THE LOTS

In total 72 MHz of DCS-1800 spectrum and 10 MHz of E-GSM spectrum was auctioned. There were 2 big lots (A and B), each consisting of 5 MHz of E-GSM-spectrum and 15 MHz of DCS-1800 spectrum. (15 MHz is the equivalent of 75 channels (frequencies).) In addition, there were 16 small lots that had in total 42 MHz of spectrum. Not all frequencies are identical however. The same frequencies have been reserved for DCS-1800 in Belgium, Germany and the Netherlands, but the operators using these frequencies differ in different countries and this might give rise to problems (of interference) in a border area. To prevent these difficulties, agreements have been made concerning which country (-operator) has priority in case of simultaneous use. Hence, there are 4 different types of frequencies: those for which one has priority both against Belgium and Germany (H), those for which one has priority only against Belgium (B), those for which one has priority only against Germany (G) and those that do not have priority (N). Table 1 provides details.

Lot	F	H	B	G	N	MHz
A	75	25	12	12	26	15
B	75	25	12	12	26	15
1	13	6	0	0	7	2.6
2	12	12	0	0	0	2.4
3	13	0	0	0	13	2.6
4	12	0	12	0	0	2.4
5	13	7	0	6	0	2.6
6	12	0	0	6	6	2.4
7	13	6	0	0	7	2.6
8	12	12	0	0	0	2.4
9	13	0	0	0	13	2.6
10	12	0	12	0	0	2.4
11	13	7	0	6	0	2.6
12	12	0	0	6	6	2.4
13	13	6	0	0	7	2.6
14	12	12	0	0	0	2.4
15	13	0	0	0	13	2.6
16	22	6	12	4	0	4.4

Table 1: Details about the lots (F= # frequencies, H= # frequencies with priority w.r.t. Belgium and Germany, B = # frequencies with priority w.r.t. Belgium, G = # frequencies with priority w.r.t. Germany, N = # frequencies without priority)

3.2. PLAYERS IN THE AUCTION

There were 7 players in the auction. The two incumbent GSM-operators, KPN and Libertel, were eligible to bid on all small lots. In addition, there were 5 new entrants: Telfort (a joint venture of BT and the Dutch railroads, that also has a licence to operate a fixed telephony network), Federa (a joint venture of France Telecom, Deutsche Telekom and 2 Dutch banks: ABN/AMRO and RABO), Orange/Veba (a consortium of the British mobile operator Orange and the German mobile operator Veba), Airtouch (a US baby bell) and TeleDanmark. France Telecom was already active on the Dutch market, in 1997 it had bought the largest Dutch cable network (Casema) when KPN had been forced to divest this. TeleDanmark was already active in Belgium, it is a shareholder in Belgacom (15 percent), together with (among others) Ameritech (20 percent), the Chicago based baby bell.

3.3. THE AUCTION RULES

The auction rules distinguish between incumbents and entrants. Incumbents (KPN and Libertel) are not allowed to bid on the lots A and B, but they are allowed to bid on all of the lots from 1, ..., 16. Each newcomer is allowed to bid on any of the lots for which he has paid a deposit. (All players had paid deposits for all of the lots for which they were eligible.)

The auction involves multiple rounds of simultaneous bidding. The minimum bid in round 1 is 0 for each lot.

At the beginning of round $t + 1$, each player receives information about what happened in round t. Specifically, for each lot it is revealed how many bids there were in the previous round, what was the highest bid (rounded to the nearest multiple of 100,000 (resp. 10,000) for the large (resp. small) lots and how many bids were highest. For those lots for which there was no activity in the previous round, the information from the round before is carried forward, i.e. each bidder is reminded of the highest bid that has been made up to now for each lot. In addition, each bidder is *privately* informed about on which lots he is standing highest.

At the beginning of each round, each player also receives information that is relevant for that round. Specifically, each player gets to hear the minimum bid that has to be made on each lot and the number of parties that is eligible to bid on that lot. The minimum bid is determined by the auctioneer. It is equal to the present (non rounded) highest bid for the lot plus an increment which lies in the range of 0 to 10 percent of that bid. A player that is eligible to bid, can bid on as many lots that he wants, with the exception that a bidder who is standing high on A (resp. B) is not allowed to bid on B (resp. A). Who is eligible to bid is determined according to the following rules:

1. If player i bids in round t, then i is eligible to bid in round $t + 1$;
2. If, when entering round t, player i is having the highest bid on lot k, and player i is overbid on k during round t, then i is eligible to bid in round $t + 1$;

The auction continues until there is a round in which no bids are made. The lots are then allocated to the bidders that are standing high at that point in time and these pay their bids. The auctioneer has the discretionary power to announce the last round if, in his opinion, the auction is taking too long. In such an announced last round, the parties that are eligible to bid are, in addition to those determined according to i) and ii) above, all those parties that are standing high when entering this announced last round. This latter rule did not have to be involved in the actual auction.

Besides the asymmetric treatment of incumbents and entrants, we may note the following differences with the auction design that was used in the US PCS-auctions:

1. In the Dutch case, there is no equivalent to the US-activity rule: a party that is interested in lot A or B is not forced to be active on a combination of lots of a similar size, he can decide to remain active only on one of the smaller lots. The absence of such an activity rule makes "hiding in the grass" a more attractive strategy, which might prolong the auction considerably.

2. In the Dutch case, a player is not allowed to withdraw his bid. As a consequence, the possibility of inefficient lock-in is more likely and bidding on the smaller lots in order to aggregate a nationwide network is less attractive.

3. In the Dutch auction, there is no common knowledge about who is standing high on which lots in a round. Each player is only privately informed about where he is standing high. However, it is not forbidden for a player to inform the others where he is standing high.

3.4. AUCTION OUTCOME

The following Table 2 specifies which bidders acquired which lots, and what price was paid. A and B licences were awarded to Federa (Dfl. 600 million) and Telfort (Dfl. 545 million), while KPN acquired 7 small lots; Libertel and Orange/Veba, 2 lots apiece; TeleDanmark 4; and Telfort 1. The State came out ahead by 1,835 million guilders.

Lot	Winning bidder	Price (Dfl x million)
A	Federa	600
B	Telfort	545
1	Libertel	40.4
2	KPN	40.2
3	Orange	38.0
4	Telfort	40.5
5	KPN	43.0
6	TeleDanmark	41.1
7	KPN	40.4
8	KPN	39.1
9	Orange	46.5
10	TeleDanmark	41.25
11	KPN	42.98
12	TeleDanmark	39.9
13	KPN	39.9
14	KPN	40.5
15	Libertel	45.5
16	TeleDanmark	71.5

Table 2: Lots acquired and price paid by the winning bidders

4. Analysis and evaluation

As Mr. Zalm, the Finance Minister, remarked afterwards, the outcome was indeed 'a nice, tidy sum'. The government, however, had consistently emphasized that the proceeds were only a secondary goal, and that the main purpose behind introducing the auction mechanism had been to create a quick, transparent process that would lead to an efficient outcome. In this section we address whether this goal was achieved. We first analyze the auction outcome; in a second subsection, we focus on some peculiarities observed during the course of the game.

4.1. THE AUCTION OUTCOME

Was this goal of efficiency really achieved? A more detailed look at the
outcome casts some doubts. First of all note that 'The Law of One Price'
does not hold: identical goods were not being sold for the same price -
something that would indeed be the case in a perfect market. Although the
technical specifications for lots 3, 9 and 15 were identical, for example, they
were sold for Dfl. 38, 46.5 and 45.5 million respectively - yielding a spread
in prices of about 20 percent around the mean. (For the other groups of
two and three identical lots, the difference was not as great.)

Another interesting feature was the price paid for each frequency. This
varied from Dfl. 2.92 million (for lot 3) to 3.58 million (for lot 9). The
standard deviation was 0.17 with a mean of 3.29 - in other words, a spread
of 5 percent. Recall, however, that different frequencies are imbued with
different rights, so that the difference in rights should be reflected in price.
By using a simple regression, the implicit price for each of these rights can
be revealed. If we limit the regression to lots 1 to 16, we find that:

$$P = \quad 3.22\,H + \quad 3.40\,G + \quad 3.34\,B + \quad 3.26\,N \quad (R^2=0.93)$$
$$\quad\quad (32.8) \quad\quad (16.6) \quad\quad (28.6) \quad\quad (35.8)$$

(with t-values indicated between parentheses).

Surprisingly, the H-frequencies (i.e. those that offer the most rights)
have the lowest price. The differences are not significant, however, and
the hypothesis that all coefficients are equal cannot be rejected (p value
$= 0.74$). Still, we believe that the behavior of the large incumbent, KPN,
demonstrates that H-frequencies do have higher value. For example, while
KPN bought approximately 1/3 of all frequencies in the small lots, of all
74 H-frequencies available in these smaller lots, KPN bought 62! Apparently, KPN believed these frequencies to constitute value for money. Even
more importantly, the regression equation illuminates a fundamental price
difference between lots A and B and the 16 small ones: on the basis of the
prices paid for the small lots, the predicted value of lots A and B is 246.4
million (with a 95% confidence interval of (236, 256)). This predicted value
differs significantly from the realized prices of 545 million and 600 million,
respectively. The difference in price can be seen in yet another way: the
combination of lots 3, 4, 8, 11, 12 and 13 yielded exactly the same capacity
(in regard to the DCS band) as A and B but cost only 240.3 million.

How can this price difference be explained? One possible explanation
may lie in the fact that the E-GSM part of the A and B licences has not yet
been taken into consideration. Before doing so, however, it is worth asking
whether the E-GSM part represents any real value: a bidder wanting to utilize both DCS-1800 and E-GSM would have to use special mobile telephones

which are not yet on the market and may not even be available within the next two years. Within that time, the bidders would already have rolled out their DCS-1800 network and would no longer need E-GSM. E-GSM, therefore, is of little value. The lower limit is zero, while an upper limit is obtained by making 1 E-GSM frequency equal to 1 DCS-1800 frequency. Since A and B each have 100 frequencies, this means that the maximum value of A and B can be estimated at 329 million which is still less than 55

A better explanation for the observed price difference lies in the auction rules, in particular the lack of transparency during the auction process and the fact that bids cannot be withdrawn. Given the inability to withdraw a bid, the cost of the lots that a bidder at present is standing high on constitute a "sunk cost", so that the bidding on the small lots becomes a war of attrition. In such a situation, the highest bid wins, but each bidder, also a loser, must pay the price of his bid. The simplest example of this is the 'dollar auction', introduced in Shubik (1971), in which one dollar is auctioned: after bidder A has bid $ 0.60 and bidder B has bid $ 0.70, bidder A must choose between accepting his loss or taking the risk of bidding even higher. The temptation is to bid higher and the lot is sold for substantially more than its true value. Experiments show that such wars of attrition commonly induce inflated prices, and specific experiments with the Federa team carried out before the auction showed this to be the case here. Players expecting such a price war may rationally decide not to bid on the smaller lots.

The endurance required of bidders is further intensified because bidders are unable to see *which* of the other bidders has made the highest bid. If a bidder decides to try to form a network that will cover the country by means of accumulating small lots, he will have to compete with all possible opponents; he cannot localize the competition in order to concentrate efforts.

The inability to withdraw bids, also implies that a bidder might get stuck with a few single lots that will turn out to be worthless (this is actually the situation in which Orange/Veba found itself). Finally, given budget constraints and the impossibility to withdraw bids, combined bidding on the large and small lots is a risky business. Every sum committed to the small lots is unavailable for bidding on the large lots, and it is quite possible that Orange/Veba lost out to Telfort and Federa precisely because it had spread itself too thin. Indeed, after the auction it was revealed that parties were willing to go to the limit of Dfl. 600 mln to obtain a licence. With this budget, Orange/Veba could not overbid Telfort on lot B as the minimum required bid was 570 and the firm had committed 84.5 on lots 3 and 9.

All these reasons indicate that bidding for small lots was not deemed attractive. In practice, bidding activity during the auction concentrated on the large lots, and no bidding activity took place on lots 1 to 16 for a long time before the auction closed. The low level of bidding for lots 1 to 16 provided KPN and Libertel with a clear advantage, where the original idea was that these bidders should be discriminated against.

In all probability, the specific auction design and the resulting avoidance of small lots by entrants has also caused the auction outcome to be inefficient. First of all, for Telfort lot 4 is redundant as lot B alone offers enough capacity, hence, lot 4 does not offer much value for this player. Secondly, the two lots that Orange/Veba acquired are insufficient to roll out a network, hence, an assignment of two lots to this party is not efficient. Before it be concluded that this party has temporarily squandered Dfl. 85 million, it should be realized that the rules allowed Orange/Veba to resell the licenses, subject to permission by the State. These rules state that reselling to other active players is not possible, but that permission might be granted if the sale is to new parties. After the auction TeleDanmark has set up a new company, currently active under the name "Ben", that has bought the licenses of both TeleDanmark and Orange/Veba. Other parties have issued a complaint that it was not justified for the Minister to give permission. In any case, the fact that resale took place is evidence of inefficiency of the auction outcome. We note that the Orange/Veba behavior appears slightly difficult to explain: given that they were willing to bid on small lots, why did they fail to continue their bidding on 1 to 16 when they realized that A and B were unobtainable? If they could have accumulated additional lots for the stated closing prices, they would have been able to put together a national network (for Dfl. 288 million if they had purchased lots 1, 2, 6, 7 and 10). Perhaps Orange/Veba also believed this strategy to be too risky: after all, the other bidders would not just stand around doing nothing, but would probably retract their lots and drive up prices.

As a third indication of inefficiency, we note that it is not clear that TeleDanmark was, after Telfort and Federa, the player with the third highest value. There were at least three parties that were willing to pay more than Dfl. 550 million for a national network, and one of these (Orange/Veba) was eliminated. If TeleDanmark valued the license at less than Dfl. 550, the third licence was thus awarded inefficiently. In any case, a different auction design, or a different packaging of the licenses, might have led to three national networks being sold for at least Dfl. 550 million each, hence the State could probably have generated more revenue.

What must be concluded, then, is that the auction outcome was not efficient and that the State could have obtained higher revenues. An alternative auction design, like, say, the model used in the USA for mobile telecommunications or a combinatorial auction, could have eliminated the problems which emerged. The American system of auctioning has three advantages over the Dutch system. Firstly, there is an activity rule that forces bidders to remain active and which accelerates the course of the auction. Secondly, bidders can withdraw their bid when they see that they are no longer able to create an efficient accumulation of lots, and this leads to more aggressive bidding. Finally, and most importantly, bidders in the American auction system have common knowledge about the state of the auction at any time, i.e. everyone knows who is currently standing the highest on any given lot. The availability of such information makes the situation transparent to participants and leads to both higher bids and more efficient allocation.

Obviously, the interesting question is why an alternative design, such as the American auction system, was not used in the Netherlands. We have already indicated the answer in the previous section: not enough time was devoted to thinking through the intricacies involved, probably because it was not realized how crucial the details in the game rules can be. Expert opinion was sought when it was not really needed, when it became more crucial, after the idea of a single new national licence was abandoned, it was not sought again.

The Minister tried to justify the choice by brushing aside the criticism and claiming that the provision of information would encourage the formation of bidder cartels, with a detrimental effect on the outcome. For example, see the remarks of Minister Jorritsma-Lebbink during the discussion on the law in the first chamber of parliament, November 25, 1997. (Verslag EK, p. 8-355). We find this rather unconvincing, given that the parties were cooped up together in the same hotel for two weeks with plenty of opportunity to conspire outside the 9-5 hours that the auction was running.

4.2. THE AUCTION PROCESS

I was asked by the Federa team to provide them with game theoretic advise on how to play in the auction. Together with other researchers from Tilburg (Sander Onderstal and Henk op den Brouw, in particular) we analyzed simple models, did simulations and performed experiments. Later on, the group of prof. Selten (Bonn) was called in for the experimental sessions as well. These mock auctions showed that bidding on the smaller lots was quite a risky business: in several experimental sessions, the smaller lots were sold at a much higher price per frequency than the large lots. The reason is the so called "exposure problem" that was already referred to above. See also Bykowski et al (1998). Similarly, the experiments and simulations showed the riskiness of bidding on both small and large lots. Having seen the risks involved, the Federa team decided that it would bid only on the large lots. Apparently, Telfort had reached a similar conclusion as it also bid exclusively on the large lots.

The next Table gives the bidding activity (average number of bids per lot) during consecutive stages of the auction. As one can see, at the start there was activity on both small and large lots. When large lots became relatively more extensive, bidding intensity on them decreased and during the periods 41-50 there was bidding only on small lots. This drove up the prices of the smaller lots which induced bidding again on large lots. The process repeated itself with no bidding on large lots during periods 71-90. After round 98, there was no more activity on the small lots.

Round	Large	Small
1-10	1.2	0.3
11-20	1.4	0.15
21-30	1.0	0.19
31-40	0.75	0.20
41-50	0	0.47
51-60	0.65	0.21
61-70	0.25	0.34
71-80	0	0.32
81-90	0	0.22
91-100	0.35	0.12
101-136	0.5	0

Table 3: Bid activity

As bidders typically made bids at the minimum required level set by the auctioneer, the auction revenue increased at a more or less constant rate. The following figure displays the evolution of revenue per frequency during the auction. We see that, throughout the auction, smaller lots were cheaper but that the difference was small around round 90. At that point, bidding intensified again on the large lots and, as already seen above, the final per unit price was much larger for the larger lots than for the smaller ones.

REVENUE PER FREQUENCY

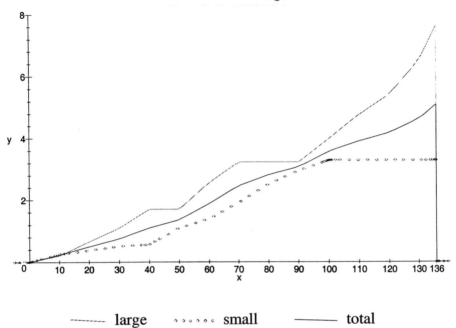

------- large   ◦ ◦ ◦ ◦ ◦ ◦ small   ——— total

From round 98 onwards, there was bidding only on large lots and only 3 parties (Federa, Orange/Veba and Telfort) were eligible to bid. At that time, Airtouch had already left the auction. It had made a jump bid (bidding more than the required minimum) of 225 on B in round 65, when this bid was overbid in round 95, it announced its withdrawal. The other parties were not active and were not activated by the bidding parties. During the auction, rumor had it that Libertel regretted this situation. According to newspaper reports, that party had made a mistake during round 98 making it ineligible to bid in future rounds. (Interestingly, round 98 took place on "Carnival Monday", a day of celebrations and drinking in the south of the Netherlands, where Libertel is based.) Bidding from round 99 till almost the very end of the auction was very boring with the current "outsider" just bidding the lowest bid required (the minimum set by the auctioneer) to replace an "insider". This pattern was broken only in round 132 when Telfort first made a slight jump bid. The next Table describes the end of the auction (Mi denotes the minimum required bid on lot i, Bi the bid, and B denotes the bidder: O = Orange, T = Telfort, F = Federa).

Round	M_a	M_b	B_a	B_b	B
132	500	510	505		T
133	520	510		510	F
134	520	530	540		O
135	560	530		545	T
136	560	570	600		F
137	630	570	- - -	- - -	- - -

Table 4: Endgame

Note the jump bid of Federa in round 136 in particular. Apparently, Federa had a budget of Dfl. 600 mln. Given this, it did not make sense to bid less than the budget: it would have increased the probability of being overbid while the budget would not have been sufficient to raise the bid. In that case, the auction would have been lost with the principals of the bidding team members having as a possible argument that this was because the team had not gone to the limit. From the agents' perspective, going to the limit was clearly the best thing to do.

As information was scarce during the auction, each player having private information, one could expect signaling to occur during the auction. In the initial phase of the auction, players signaled by using special combinations of digits and by bidding on consecutive lots. For example, in round 46, the bids on lots 1-4 ended on .21, while those on 9-14 ended on .25. The clearest signals were sent by TeleDanmark. In round 37 this bidder sent around a memo to the other bidders stating "please note that TeleDanmark was highest on the lots 1 through 6 with prices of 9.21, 9.21, 9.21, 9.21, 9.21, and 7.61". In rounds 38-41 this bidder was not overbid on these lots and it sent a memo saying "please note that TeleDanmark did not bid in this round". In following rounds, occasionally lots would be taken away from TeleDanmark with the bidder taking them back and it notifying others with memo's of the above type. Clearly, TeleDanmark was trying to scare others away from six of the small lots by signaling its intentions. It, however, was not successful with this strategy and, by round 66, it stopped the signaling. From a theoretical point of view, it is interesting to know whether such signaling is advantageous, should the other players have reacted to it?

During the process, the Federa team was trying to reconstruct who was standing high on which of the small lots. Not bidding on the small lots itself, many combinations were possible and, in the end, the attempt to construct the state of the auction was not successful: there were too many degrees of freedom left after the signaling had stopped. Also in this respect, an incumbent like KPN, that was bidding extensively on small lots, had a considerable advantage.

5. Conclusion

The case described above is interesting for a couple of reasons:

1. The gaming that took place before the auction. Well aware of the fact that EU-directives force the minister to allocate all available spectrum, the incumbent, KPN, released the spectrum that was under its control at strategic points in time and thereby, more or less, made the government its hostage. The policy was successful in delaying the awarding of new licenses and helped in having an auction design that favored the incumbent .

2. The interaction between the Dutch government and the European Commission. It remains interesting why the Commission ruled differently in this case than in the previous GSM cases in Belgium, Italy and Spain. Also interesting is the fact that KPN was successful as a free rider on Libertel. The Commission argued that Libertel could not be excluded and that a levy could not be imposed on Libertel; in the end also KPN was not excluded and neither did it have to pay for the GSM license that it had gotten for free.

3. The auction game itself. In essential aspects, the rules are different from those that were used in the US and a comparison is instructive. We discussed some of this above, but clearly much more remains to be said, in particular theoretical and experimental analyses are invited: how does the packaging of the lots influence the outcome?, what are the consequences of allowing bid withdrawal?, is the exposure problem more serious when the state of the auction is not common knowledge?, is it desirable to signal? Some of this work has already begun, see, for example, Onderstal (1999) in which a simple theoretical model is given illustrating that a design as in the Dutch case might indeed result in lower seller revenue than what could be obtained.

4. The different role that economists in the Netherlands played in the process as compared to the role played by American academics.

My overall conclusion is that the auction format that was used in the Netherlands was not "optimal". Not only does it not generate maximal revenue for the government, neither does it guarantee an efficient allocation.

In addition, it favored the incumbents. It would have been better to give more information to parties during the auction or to simply use the same auction format as was used in the US PCS-auctions. The main argument that was given before the auction for why the US-auction format was not used was that collusion should be prevented and that collusion would be too easy in that design. I don't find this argument to be convincing. Contacts between parties were not forbidden and, in response to the question by one of the parties of whether it could reveal information outside of the auction, it was answered that this was not forbidden. The bidding teams were cooped in the same hotel for 3 weeks so that there were plenty of opportunities for colluding outside of the game for those that desired to do so.

After I had published my critique on the auction in Dutch (Van Damme, 1998), questions were asked in parliament and the Secretary of State in charge addressed the above criticisms in a letter to parliament dated January 11, 1999. She argues that revenue maximization was not a goal and that any critique that revenue could have been higher is beside the point. More importantly, she disagrees with the conclusion that the auction outcome was inefficient. In particular, she argues that there is no evidence that the law of one price does not hold since also the E-GSM frequencies have value. As we have seen above, even taking their maximum possible value into account, a substantial price gap remains between the large and the small lots. Furthermore, we have given other indications of inefficiency (resale and Orange willing to pay more than what TeleDanmark paid) that the Secretary does not try to refute. The Secretary also remains unconvinced that a different design could have done better: an activity rule would limit the flexibility of bidders too much, bid withdrawals would make current bids uninformative and providing more information could have led to strategic bidding and to collusion. She concludes:

"The chosen auction model, simultaneous multi-round with many opportunities to switch between lots, guaranteed fine competition and proved adequate. The auction was quick and successful. The goal of the auction, to allocate scarce radio frequencies for mobile telephony in a transparent way has been reached".

I think the final word has not yet been said about this auction.

References

Bykowsky M., M. Olson en A. Schram. "Veiling van etherfrequenties." *Economisch Statistische Berichten*, 1-3-1995, 201-205.
Bykowsky, M., R. Cull and J. Ledyard. "Mutually destructive bidding: the FCC auction design problem." Social Science Working Paper 916 Cal. Tech. 1998.

Cramton, P. "Money out of thin air: the nationwide narrowband PCS Auction." *J. Economics and Management Strategy* **4** (1995) 267-345

Cramton, P. "The PCS spectrum auctions: An early assessment." *J. Econ. Manag. Strat.* **6** (1997) 431-497

Damme, E.E.C. van. "Aanbesteding en veilingmechanismen. Economische theorie en toepassingen." Research Series on Competition. Ministry of Economic Affairs, January 1997a.

Damme, E.E.C. van. "Tien misverstanden over veilingen." *Economisch Statistische Berichten,* January 8, 1997b.

Damme, E.E.C. van. "Veilen in de praktijk: mobiele telefonie frequenties." *Economisch Statistische Berichten,* April 10, 1998.

DDV Telecommunications consultancy "Het veilen van frequenties: verslag van de consultatieronden". (Inclusief binnengekomen schriftelijke reacties). (1996)

European Commission: "Directive 96/2/EG of January 16, 1996 concerning a change in directive 90/388/EEG with respect to mobile and personal communication." PB L20, 26/1/96.

European Commission: "Green paper on mobile and personal communication." (...)

HDTP: "Frequentiebeleid in Nederland; communicerende golven." Ministerie van Verkeer en Waterstaat. Groningen, 1993.

Keuter, A. and L. Nett: "Ermes-auction in Germany". *Telecommunications Policy* **21** (1997) 297-307

McAfee, R.P. and J. McMillan. "Analyzing the airwaves auction." *J. Economic Perspectives* 10 (1996) 159-175.

McMillan, J. "Selling spectrum rights." *J. Economic Perspectives* **8** (1994) 145-162.

McMillan, J. "Why auction the spectrum?" *Telecommunications Policy* **19** (1995) 191-199.

Milgrom, P. "Auction theory for privatization." Manuscript of a book to be published by Cambridge University Press, Stanford University, 1996.

Milgrom, P. "Game Theory and the spectrum auctions." *Eur. Econ. Review* **42** (1998) 771-778.

Olsen, M., A. Schram and F. van Winden: "De veiling van etherfrequenties door de overheid; een verslag van de bevindingen van CREED". Report dated September 30, 1996.

Onderstal, S. "The racket auction." Discussion paper, CentER ,Tilburg University, 1999.

Shubik, M. "The dollar auction: a paradox in non-cooperative behavior and escalation." *J. Conflict Resolution* **15** (1971) 109-111.

Tweede Kamer: "Wijziging van de wet op de telecommunicatievoorziening in verband met de invoering van het veilen van schaarse frequenties voor systemen van digitale mobiele telecommunicatie." Kamerstuk 25171, nr. 1,2, 1996-1997.

CHAPTER 6

BIRD'S TREE ALLOCATIONS REVISITED *

VINCENT FELTKAMP (vincent.feltkamp@bigfoot.com)
RNO-RSM, Statistics Netherlands,
P.O. Box 4481, 6401 CZ Heerlen, The Netherlands

STEF TIJS (s.h.tijs@kub.nl)
CentER and Econometrics Department, Tilburg University,
P.O. Box 90153, 5000 LE Tilburg, The Netherlands

SHIGEO MUTO (muto@vs.valdes.titech.ac.jp)[†]
Dept of Value and Decision Science,
Graduate School of Decision Science and Technology,
Tokyo Institute of Technology,
2-12-1 Oh Okayama, Meguro-ku, Tokyo 152-8552, Japan.

Abstract. Minimum cost spanning tree (mcst) construction and cost allocation problems have been studied extensively in the literature, though usually not together. Bird (1976) proposes an allocation rule of which Granot and Huberman (1981) prove that it lies in the core of the associated mcst game. We show that the problems of finding an mcst and allocating its cost can be integrated. Furthermore, we provide an axiomatic characterization of the set of all Bird's tree allocations using consistency, converse consistency, non-emptyness and efficiency, and give a strategic form game of which the set of Nash equilibria contains Bird's tree allocations.

1. Introduction

Consider a group of villages, each of which needs to be connected directly or via other villages to a source. Such a connection needs costly links.

---

* Inspiring discussions with Peter Borm, Michael Maschler, Gert-Jan Otten, Jos Potters, Hans Reijnierse and Oscar Volij about the subject of the paper are gratefully acknowledged.

† This author wishes to acknowledge the Canon Foundation in Europe Visiting Research Fellowship which made his visit to CentER possible

Each village could connect itself directly to the source, but by cooperating costs might be reduced. This cost minimization problem is an old problem in Operations Research, and Borůvka (1926) came up with algorithms to construct a tree connecting every village to the source with minimal total cost. Later, Kruskal (1956), Prim (1957) and Dijkstra (1959) found similar algorithms. A historic overview of this minimization problem can be found in Graham and Hell (1985).

However, finding a minimal cost spanning tree (mcst) is only part of the problem: if the cost of this tree has to be borne by the villages, then a cost allocation problem has to be addressed as well. Claus and Kleitman (1973) introduced this cost allocation problem, whereupon Bird (1976) treated this problem with game-theoretic methods and for each minimum cost spanning tree proposed a cost allocation associated with it. We call the allocations yielded by this rule *Bird's tree allocations*. As there can be more than one mcst for a given problem, Bird's rule can yield more than one allocation. However, generically, there is only one mcst and then this rule yields a unique allocation.

Granot and Huberman (1981) proved that Bird's tree allocations are extremal points of the core of the associated minimum cost spanning tree game. This game is defined as follows: the players are the villages and the worth of a coalition is the minimal cost of connecting this coalition to the source via links between members of this coalition. Not being satisfied with only one extremal point of the core, Granot and Huberman then provide the weak and strong demand operations, which yield more core elements when applied to Bird's tree allocations. Aarts (1992) found other extreme points of the core in case the mcst problem has an mcst that is a *chain*, i.e. a tree with only two leaves. Kuipers (1993) investigated the core of information graph games. These are games arising from mcst situations in which the costs of links are either one or zero.

Granot and Huberman's reason for looking at other core allocations than those yielded by Bird's rule, is that although core elements are stable against defection by subcoalitions, an extremal point of the core discriminates against some players. For example, Bird's tree allocations discriminate against the players closest to the root. Granot and Huberman's demand operations remedy this problem by allowing players to demand contributions from players that 'need' them.

In this paper we provide two new views of Bird's tree allocations: an axiomatic characterization of the set of Bird's tree allocations, and a non-cooperative game, in which Bird's tree allocations correspond to Nash equilibria.

We treat the Operations Research problem and the cost allocation problem simultaneously. One reason is that they are two sides of the same problem, and solving one side gives insight into the other side. For example, examining Bird's tree allocation rule for minimum cost spanning tree problems, one sees that it is intimately linked to the algorithm for finding minimum cost spanning trees that is described in Prim (1957) and Dijkstra (1959). Hence, we integrate Bird's tree allocation rule into Prim and Dijkstra's algorithm. Doing this suggests finding allocation rules that correspond to the other algorithms for finding minimum cost spanning trees, viz. the algorithm of Kruskal (1956), and the decentralized algorithm that was first described in Borůvka (1926). This will be done in Feltkamp, Tijs and Muto (1994b). Furthermore, when axiomatizing cost allocation solutions that associate a cost allocation to each minimal extension knowing which extension a particular allocation is associated with is useful.

The outline of this paper is as follows.

Section 2 presents minimum cost spanning tree problems and Bird's rule. Instead of solving the Operations Research and cost allocation problems consecutively, they are integrated: the cost of a link in an mcst is allocated at the same moment the link is constructed in the process of tree formation. Section 3 characterizes the set of trees and corresponding allocations yielded by Bird's rule axiomatically, using non-emptyness, efficiency, consistency and converse consistency. Section 4 presents a non-cooperative game, in which a strategy of a player consists of choosing how much to contribute to the cost of his incident links. Bird's tree allocations will turn out to be Nash equilibria of this game. Section 5 concludes.

Preliminaries and notations

We refer to any elementary textbook on graph theory for an understanding of graph theory, but recall some definitions to show the notational conventions. A graph $< V, E >$ consists of a set V of vertices and a set E of edges. An edge e incident with two vertices i and j is identified with $\{i, j\}$[1]. For a graph $< V, E >$ and a set $W \subseteq V$,

$$E(W) := \{e \in E \mid e \subseteq W\}$$

is the set of edges linking two vertices in W. For a set $E' \subseteq E$,

$$V(E') := \{v \in V \mid \text{there exists an edge } e \in E' \text{ with } v \in e\}$$

is the set of vertices incident with E'.

The complete graph on a vertex set V is the graph $K_V = < V, E_V >$, where

$$E_V := \{\{v, w\} \mid v, w \in V \text{ and } v \neq w\}.$$

---

[1] Because we do not consider multigraphs: two vertices are connected by at most one edge.

A *path* from i to j in a graph $< V, E >$ is a sequence $(i = i_0, i_1, \ldots, i_k = j)$ of vertices such that for all $l \leq k$, the edge $\{i_{l-1}, i_l\}$ lies in E. A *cycle* is a path of which the begin-point coincides with the end-point. Two vertices $i, j \in V$ are *connected* in a graph $< V, E >$ if there is a path from i to j in $< V, E >$. A subset W of V is *connected* in $< V, E >$ if every two vertices $i, j \in W$ are connected in the subgraph $< W, E(W) >$. A connected set W is a *connected component* of the graph $< V, E >$ if it is connected and no superset of W is connected. A *connected graph* is a graph $< V, E >$ with V connected in $< V, E >$. A *tree* is a connected graph that contains no cycles. A *leaf* of a tree is a vertex that is incident to only one edge of the tree.

The cardinality of a set S will be denoted by $|S|$.

With many economic situations in which costs have to be divided one can associate a *cost game* (N, c) consisting of a finite set N of players, and a *characteristic function* $c : 2^N \to R$, with $c(\emptyset) = 0$. Here $c(S)$ represents the minimal cost for coalition S if it secedes, i.e. if people of S cooperate and can not count upon help from people outside S.

The economic situations in the sequel involve a set N of users of a source $*$. For a coalition $S \subseteq N$, we denote $S \cup \{*\}$ by S^*. Furthermore, for a vector $x \in R^N$ and a player $i \in N$, we denote x^{-i} the restriction of x to $N \setminus \{i\}$.

The *core* of a cost game (N, c), is defined by

$$\text{Core}(c) = \{x \in R^N \mid \sum_{i \in N} x_i = c(N) \text{ and } \sum_{i \in S} x_i \leq c(S) \text{ for all } S \subseteq N\}.$$

2. Mcst problems and Bird's tree allocation rule

Formally, a *minimum cost spanning tree (mcst) problem* $< N, *, w >$ consists of a finite group N of agents, each of whom wants to be connected to a common source, denoted by $*$. The non-negative cost of constructing a link $\{i, j\}$ between the vertices i and j in $N^* \equiv N \cup \{*\}$ is denoted by $w(i, j)$. Because of these costs, agents have an incentive to cooperate, and to construct a minimal cost graph that connects them all to the source. If a cycle appears in such a *minimum cost spanning graph*, at least one edge in this cycle can be eliminated, to yield another minimum cost spanning graph, with less cycles. Hence, there are minimum cost spanning graphs that contain no cycles at all, i.e. that are trees. This explains the name of the problem.

Prim (1957) and Dijkstra (1959) proposed the following algorithm to find a minimum cost spanning tree given an mcst problem.

ALGORITHM 2.1. (Prim and Dijkstra).
  input :   an mcst problem $\mathcal{T} \equiv < N, *, w >$
  output : the edge set T of a minimum cost spanning tree

1. Choose a vertex $v \in N^*$ as *root*.
2. Initialize $T = \emptyset$.
3. Find a minimal cost edge $e \in E_{N^*} \setminus T$ incident to $\{v\} \cup N^*(T)$ such that joining e to T does not introduce a cycle.
4. Join e to T.
5. If not all vertices are connected to the root in the graph $< N^*, T >$, go back to step 3.

Prim and Dijkstra then prove that any tree resulting from the algorithm is an mcst. Note that by varying between the possible edges in step 3, this algorithm can construct all minimum cost spanning trees of this mcst problem $\mathcal{T}$.

A problem related to such a minimization problem is how to allocate the cost of the edges of a minimum cost spanning tree among the agents (users of the source) in a reasonable way. Bird (1976) associated the following transferable utility mcst-game $(N, c^{\mathcal{T}})$ to an mcst problem $\mathcal{T}$. The players are the agents and the worth $c^{\mathcal{T}}(S)$ of a coalition S is the minimal cost of a spanning tree on $S^* := S \cup \{*\}$. In formula,

$$c^{\mathcal{T}}(S) = \min\{\sum_{e \in T} w(e) \mid T \subseteq E_S \text{ and } < S^*, T > \text{ is a spanning tree}\}$$

for all $S \subseteq N$. Bird also proposed a cost allocation rule for the mcst problem, which he calls the *tree allocation* rule, because it associates a cost allocation to every mcst of the mcst problem. Granot and Huberman (1981) proved that Bird's tree allocation rule yields extreme points of the core of the mcst game. Given an mcst problem $< N, *, w >$ and a mcst $< N^*, T >$ for the grand coalition, Bird's tree allocation β^T is constructed by assigning to a player $i \in N$ the cost of the first edge on the unique path in the tree $< N^*, T >$ from player i to the source $*$. In fact, this allocation is intimately linked with the Prim-Dijkstra algorithm: the tree $< N^*, T >$ and the allocation β^T can be constructed together by choosing the source as root and allocating the cost of the edge added at a certain step to the person that this edge newly connects to the source.

If the mcst problem contains two or more edges with the same weight, there might be more than one mcst, and for a particular mcst $< N^*, T >$, it can happen that there is more than one order in which Prim-Dijkstra's algorithm can choose the edges in T. Obviously, the order does not change the edge that a player has to pay according to Bird's tree allocation rule, so Bird's tree allocation β^T is independent of the *order* in which the edges of the particular tree $< N^*, T >$ are chosen. It does, however, depend on which tree is constructed. See example 2.2.

EXAMPLE 2.2.

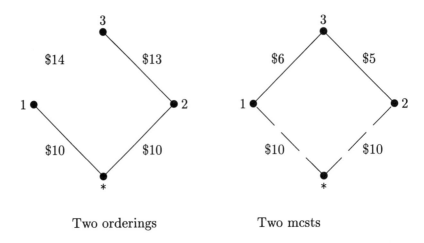

Two orderings   Two mcsts

Figure 1.   Costs that are not indicated are equal to $100.

In the problem on the left of figure 1, Prim-Dijkstra's algorithm can choose the links in the unique mcst in two orders: $(\{*,1\}, \{*,2\}, \{2,3\})$ or $(\{*,2\}, \{*,1\}, \{2,3\})$, but the unique Bird allocation equals $(10,10,13)$, and the link $\{*,1\}$ is always paid by player 1 and the link $\{*,2\}$ is always paid by player 2.

In the problem on the right of figure 1, only one of the two dashed links will be constructed. In case $\{*,1\}$ is constructed, Bird's tree allocation is $(10,5,6)$ and if $\{*,2\}$ is constructed, Bird's tree allocation is $(6,10,5)$.

3. An axiomatic characterization

In this section, we characterize the set of mcsts and their associated Bird allocations axiomatically, using non-emptyness, efficiency, consistency and converse consistency.

A *solution* of mcst problems is a function ψ assigning to every mcst problem $\mathcal{T} = <N, *, w>$, a set

$$\psi(\mathcal{T}) \subseteq \{(T,x) \mid <N^*, T> \text{ is connected and } \sum_{i \in N} x_i \geq \sum_{e \in T} w(e)\}.$$

Note that we don't ask that all these graphs $<N^*, T>$ be trees. One could suggest alternative presentations, in which a solution consists only of a set of allocations. However, the structure of the set of Bird allocations is much more clear if the trees with which the Bird allocations are associated are included in the solution.

We mention a few properties of solutions of mcst problems.

DEFINITION 3.1.

NE A solution ψ is called *non-empty* if

$$\psi(\mathcal{T}) \neq \emptyset \qquad \text{for all mcst problems } \mathcal{T}.$$

Eff ψ is *efficient* if for all mcst problems $\mathcal{T}$, all $(T, x) \in \psi(\mathcal{T})$ are *efficient*, that is, for all $(T, x) \in \psi(\mathcal{T})$, $< N^*, T >$ is a minimal cost spanning tree and

$$\sum_{i \in N} x_i = \sum_{e \in T} w(e).$$

The two properties of consistency and converse consistency use reduced mcst problems. Here, a reduced mcst problem is an mcst problem where some players have been eliminated. The idea is that solving reduced problems is easier than solving the original problem, and that the solution of the original problem should be related to the solution of the reduced problems. We only ask for this relation if one player that is a leaf is deleted, however. The idea is that a leaf is not needed by any other player to get connected to the source, so if a leaf player is missing, this should not affect the other players.

DEFINITION 3.2.   Given an mcst problem $\mathcal{T} \equiv < N, *, w >$ and a player $i \in N$, define the *reduced* mcst problem

$$\mathcal{T}^{-i} :=< N \setminus \{i\}, *, w^{-i} >,$$

where w^{-i} is w restricted to $E_{N^* \setminus \{i\}}$.

Note that the reduced problem does not depend on any fixed solution. Note also that it is indeed an mcst problem. Using reduced mcst problems, we can define consistency and converse consistency as follows.

DEFINITION 3.3.

Cons A solution ψ of mcst problems is *consistent* if for every mcst problem $\mathcal{T}$, for every $(T, x) \in \psi(\mathcal{T})$ and for every player i that is a leaf in the graph $< N^*, T >$,

$$(T^{-i}, x^{-i}) \in \psi(\mathcal{T}^{-i}),$$

where T^{-i} is obtained from T by deleting the unique edge incident to i and x^{-i} is the vector obtained from x by deleting the coordinate of player i.

CoCons A solution ψ of mcst problems is *converse consistent* if for any mcst problem $\mathcal{T}$ and for any (T, x) efficient for $\mathcal{T}$, the following is satisfied: if

$$(T^{-i}, x^{-i}) \in \psi(\mathcal{T}^{-i})$$

for some player i that is a leaf of $< N^*, T >$, then

$$(T, x) \in \psi(\mathcal{T}).$$

    The converse consistency property is motivated by the idea that if a possible efficient solution is excluded, the 'reduced' solution should also be excluded as solution in a reduced problem where a leaf has been deleted. It ensures that solutions that satisfy it are as large as possible without violating efficiency and consistency.

DEFINITION 3.4.   The Bird solution of an mcst problem $\mathcal{T}$ is the set

$$\beta(\mathcal{T}) := \{(T, \beta^T(\mathcal{T})) \mid < N^*, T > \text{ is an mcst of } \mathcal{T}\}$$

of sequences of edges of minimum cost spanning trees and the corresponding Bird tree allocations.

PROPOSITION 3.5.   The Bird solution satisfies NE, Eff, Cons and Co-Cons.

Proof : Efficiency was proven by Bird and non-emptiness is evident. In order to prove Cons, assume $(T, \beta^T) \in \beta(\mathcal{T})$. Any tree has at least two leaves, so let player $i \neq *$ be a leaf in the tree $< N^*, T >$. Define e to be the edge incident to i on the unique path in $< N^*, T >$ from i to the source. Then T^{-i} is obtained from T by deleting the edge e, $x^{-i} = \beta^{T \setminus \{e\}}$ and

$$(T^{-i}, x^{-i}) \in \beta(\mathcal{T}^{-i}).$$

    In order to prove that the Bird solution satisfies CoCons, assume that (T, x) is efficient in an mcst problem $\mathcal{T}$ and assume that player i is a leaf of $< N^*, T >$ such that

$$(T^{-i}, x^{-i}) \in \beta(\mathcal{T}^{-i}). \tag{1}$$

Define e_i to be the unique edge incident to i in T. Then $T = T^{-i} \cup \{e_i\}$ and $< N^* \setminus \{i\}, T^{-i} >$ is an mcst for the reduced mcst problem $\mathcal{T}^{-i}$. Hence efficiency of (T, x) and equation 1 imply

$$\sum_{k \in N} x_k = \sum_{e \in T} w(e) = \sum_{e \in T^{-i}} w(e) + w(e_i) = \sum_{k \in N \setminus \{i\}} x_k^{-i} + w(e_i),$$

which implies that $x_i = w(e_i)$. So $(T, x) \in \beta(\mathcal{T})$.   ∎

LEMMA 3.6.   If a solution ϕ satisfies Eff and Cons and a solution ψ satisfies NE, Eff and CoCons, then $\phi(\mathcal{T}) \subseteq \psi(\mathcal{T})$ for all mcst problems $\mathcal{T}$.

Proof : We proceed by induction on the cardinality of N. Let $|N| = 1$ and denote by e the edge between the unique player and the source. By efficiency of both solutions and non-emptiness of ψ, we obtain $\phi(\mathcal{T}) \subseteq \{(\{e\}, w(e))\} = \psi(\mathcal{T})$. Take an mcst problem $\mathcal{T}$ with $k > 1$ players, and suppose that for all mcst problems $\mathcal{T}'$ with less than k players, $\phi(\mathcal{T}') \subseteq \psi(\mathcal{T}')$. Take $(T, x) \in \phi(\mathcal{T})$ and choose a leaf $i \neq *$ of the tree $< N^*, T >$. Then by consistency of ϕ, $(T^{-i}, x^{-i}) \in \phi(\mathcal{T}^{-i}) \subseteq \psi(\mathcal{T}^{-i})$. Now (T, x) is efficient, hence $(T, x) \in \psi(\mathcal{T})$ by converse consistency of ψ.   ∎

THEOREM 3.7.   The unique solution that satisfies NE, Eff, Cons, and CoCons is the Bird solution.

Proof : The Bird solution has the properties, and if another solution has the properties, by lemma 3.6, it coincides with the Bird solution.   ∎

The properties used to characterize the Bird solution are logically independent. We show this by giving examples of solutions that satisfy three of the four properties.

EXAMPLE 3.8.   If we omit the non-emptiness property, the empty solution that assigns the empty set to every mcst problem satisfies Eff, Cons and CoCons.

EXAMPLE 3.9.   If we omit efficiency, assigning $(E_{N^*}, (a, \dots, a))$ to every mcst problem, yields a solution that satisfies the other three properties. Here E_{N^*} denotes the set of all edges of the complete graph on N^* and $a = \sum_{e \in E_{N^*}} w(e)$. Notice that there are no leaves in the complete graph, except if there is only one player, so the consistency and converse consistency properties are satisfied vacuously.

EXAMPLE 3.10. If we omit consistency, the solution that assigns to an mcst problem $< N, *, w >$ the set of all efficient outcomes

$$\{(T,x) \mid\ < N^*, T > \text{ is an mcst, } x \in R^N \text{ and } x(N) = \sum_{e \in T} w(e)\}$$

satisfies the three other properties.

For the last example, we pick a total ordering $<$ on the universe of all possible players. This is possible: usually, names of players are finite strings in some finite alphabet, which can be alphabetically ordered. Define the lexicographical order on elements of a solution to an mcst problem $< N, *, w >$ by $(T,x) \prec_L (\tilde{T}, y)$ if there exists a $k \in N$ such that $x_i = y_i$ for $i < k$ and $x_k < y_k$.

EXAMPLE 3.11. If we omit converse consistency, the solution that assigns to every mcst problem the set of lexicographically smallest elements of the Bird solution satisfies the three other properties. Clearly it is efficient and non-empty. Suppose it did not satisfy consistency. Then there would be a mcst problem $\mathcal{T} = < N, *, w >$ with a lexicographically minimal element (T, β^T) of $\beta(\mathcal{T})$ and a leaf i such that deleting i, the edge e incident to i and the coordinate β_i^T of i yields $(T^{-i}, \beta^{T^{-i}})$, which is not a lexicographically minimal element of $\beta(\mathcal{T}^{-i})$. Hence, there is an element $(\tilde{T}, x)$ of the Bird solution which is lexicographically smaller. But including e and β_i^T in this solution element yields $(\tilde{T} \cup \{e\}, (x, \beta_i^T)) \in \beta(\mathcal{T})$, which is lexicographically smaller than (T, β^T). This is a contradiction, thus consistency is satisfied.

This solution does not coincide with the Bird solution on all mcst problems, so it does not satisfy converse consistency.

4. Sustaining the Bird tree allocations by Nash equilibria

In the previous sections we studied mcst problems by means of cooperative games. In this section, we model the problems by strategic games, in which an action of a player consists of a specification of the edges to which this player will contribute, and which amounts he will contribute.

DEFINITION 4.1.

To a minimum cost spanning tree problem $< N, *, w >$, we associate the strategic game $< N, (A^i)_{i \in N}, (u_i)_{i \in N} >$ in normal form with player set

N, and in which an action $a^i = (a^i_j)_{j \in N^* \setminus \{i\}} \in A^i \equiv R_+^{N^* \setminus \{i\}}$ of a player i specifies for each other vertex j (j can be a player or the source), the non-negative amount a^i_j that player i is willing to contribute to the cost of the edge $\{i, j\}^2$. The utility that player i derives from a strategy profile $a = (a^i)_{i \in N}$ is determined in the following way. We assume that players dislike making contributions, but they absolutely have to be connected to the source. So the utility function is linear in the contributions, and a big penalty is subtracted if the player is not connected to the source in the graph resulting from the contributions of all players. More precisely, for a strategy profile a, the set C_a of edges that have been completely paid for and that will be constructed is defined as

$$C_a := \{\{i, j\} \in E_{N^*} \mid a^i_j + a^j_i \geq w(e)\}$$

and the utility of player i is defined as

$$u_i(a^1, \ldots, a^n) := \begin{cases} - \displaystyle\sum_{j \in N^* \setminus \{i\}} a^i_j & \text{if } i \text{ is connected to the source in} \\ & < N^*, C_a > \\ - \displaystyle\sum_{j \in N^* \setminus \{i\}} a^i_j - P & \text{otherwise} \end{cases}$$

where P is a large positive number ($P > \displaystyle\sum_{e \in E_{N^*}} w(e)$).

We proceed to establish a relationship between the Bird solution presented in section 2 and Nash equilibria of the above strategic mcst game.

THEOREM 4.2.   Each element $((e^1, \ldots, e^\tau), x)$ of the Bird solution of an mcst problem corresponds to a Nash equilibrium of the associated strategic mcst game, in which the strategy of a player i is to construct the first edge on the unique path from i to the source in the tree $< N^*, \{e^1, \ldots, e^\tau\} >$ and his payoff equals $-x_i$.

Proof : Let $\mathcal{T} = < N, *, w >$ be an mcst problem and let $((e^1, \ldots, e^\tau), x)$ be an element of the Bird solution $\beta(\mathcal{T})$. The corresponding strategy a^i for a player i is to contribute only to the first edge e_i that lies on the unique path from i to the source in the tree $< N^*, \{e^1, \ldots, e^\tau\} >$ and to pay the

---

^{2} Allowing players to pay edges not incident to themselves would lead to unwanted equilibria.

cost $w(e_i)$ of this edge completely. If every player plays this strategy, the resulting set C_a of constructed edges is precisely $\{e^1, \ldots, e^\tau\}$, which implies that all players are connected to the source. So the payoff to player i equals $-w(e_i)$. Hence $u_i(a) = -w(e_i) = -x_i$.

To prove that a is a Nash equilibrium, suppose that a player i deviates from a. It is clear that contributing more to the edge e_i does not improve i's payoff. If i contributes less than $w(e_i)$ to the edge e_i, then e_i is not paid completely, so it will not be constructed. Now player i wants to avoid the penalty, which is larger than $w(e_i)$, so i has to pay at least one other edge e' that connects the connected component of i in the graph $< N^*, C_a \setminus \{e_i\} >$ to the connected component of the source. Because $< N^*, \{e^1, \ldots, e^\tau\} >$ is an mcst, such an edge e' has to be at least as costly as the edge e_i. Hence i is not better off.   ∎

If the costs of all edges are positive, it can be proved that in all Nash equilibria of the strategic game a spanning tree is formed, although it does not have to be an mcst.

THEOREM 4.3.   If in an mcst problem the costs of all edges are positive, each Nash equilibrium a of the strategic mcst game specifies a spanning tree $< N^*, C_a >$ for the mcst problem, and the payoff vector is $-\beta^{C_a}$. Here, for any spanning tree $< N^*, T >$ we again denote β^T the allocation assigning to every player the cost of the first edge on the unique path from the player to the source in this tree.

Proof : Let $< N, *, w >$ be an mcst problem in which the costs of all edges is positive and let $< N, (A^i)_{i \in N}, (u_i)_{i \in N} >$ be the associated strategic mcst game. Let $a = (a^1, \ldots, a^n)$ be a Nash equilibrium and consider the set C_a of edges that have been completely paid. If a player i is not connected to the source in the graph $< N^*, C_a >$, then by deviating and using the strategy $\hat{a}$ in which he pays precisely the link $\{i, *\}$, he can improve his payoff. So in a Nash equilibrium, every player is connected to the root. Furthermore, if a cycle were present in the graph $< N^*, C_a >$, there is a player i that contributes a positive amount to an edge of this cycle. Then i can improve his payoff by not contributing to this edge. Hence $< N^*, C_a >$ is a tree.

It is clear that no edge of the constructed tree will be 'overpaid' and that no other edge will be contributed to. Furthermore, every player i contributes only to edges that lie on the path in the tree from i to the source. If this were not true, some player could reduce his contribution to an edge that does not lie on the path from himself to the source, without incurring the penalty. Now there is only one edge in the path from i to the source in the tree that i can contribute to, and that is the edge $e = \{i, j\}$ in the tree, that is the first edge on the unique path in the tree from i to the source. By an induction argument one sees that no other player contributes anything to this edge, which implies that i pays $w(e)$ alone. So $a_j^i = w(e)$ and $a_k^i = 0$ for all other $k \in N^*$, and the payoff is $u_i(a) = -w(e) = \beta^{C_a}$. ∎

Not all Nash equilibria correspond to mcsts, as is shown by the next example.

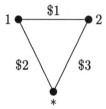

Figure 2.   A problem with a Nash equilibrium that does not correspond to an mcst.

EXAMPLE 4.4.   Consider the problem drawn in figure 2. The strategy pair where player 1 pays the edge $\{1, 2\}$ and player 2 pays $\{2, *\}$ is a Nash equilibrium[3] in the associated strategic game, but the unique mcst uses the edge $\{1, *\}$ instead of $\{2, *\}$.

We call a Nash equilibrium *total payoff maximizing* if the sum of the payoffs to the players in this equilibrium is maximal among all Nash equilibria.

THEOREM 4.5.   In an mcst problem with positive weights, the total payoff maximizing Nash equilibria correspond to mcsts.

Proof : We know that a Nash equilibrium corresponds to a tree, and that the payoffs of the players correspond to costs of edges in the tree. If the total of the payoffs is maximal, the cost of the tree is minimal, hence the tree is a mcst. ∎

Remark.   Total payoff maximizing Nash equilibria are strong Nash equilibria (private communication by Gert-Jan Otten). In a strong equilibrium, deviations by coalitions of players do not strictly improve the payoffs of all players in the coalition. Hence, theorem 4.2 can be replaced by a stronger theorem, viz.,

---

[3] Note that we do not allow a player to choose a non-incident edge, so player 2 cannot choose edge $\{1, *\}$.

THEOREM 4.6.   Each element $((e^1, \dots, e^\tau), x)$ of the Bird solution of an mcst problem corresponds to a strong Nash equilibrium of the associated strategic mcst game, in which the strategy of a player i is to construct the first edge on the unique path from i to the source in the tree $< N^*, \{e^1, \dots, e^\tau\} >$ and his payoff equals $-x_i$.

However, the non-total-payoff-maximizing equilibrium in example 4.4 is a strong Nash equilibrium. Hence, not all strong equilibria correspond to elements of the Bird solution in all strategic mcst games.

5. Conclusion

In this paper, we have reconsidered Bird's tree allocations for mcst problems, and have studied them from two points of view. First we gave an axiomatic characterization which permits us to determine whether the allocation rule is applicable. Second we introduced a non-cooperative game, of which the total payoff maximizing Nash equilibria coincide with Bird's tree allocations. For other strategic mcst games, in which a player pays (parts of) edges conditionally upon being connected in the resulting graph, see Feltkamp (1998). This however generates also non-efficient allocations as Nash equilibria.

Instead of using Bird's tree allocation rule, one could use elements of the irreducible core of an mcst problem to evaluate it. The irreducible core was also introduced by Bird (1976) and studied by Granot and Huberman (1982) and Aarts and Driessen (1993). We axiomatically characterized the irreducible core of a generalization of mcst problems in Feltkamp, Tijs and Muto (1994a), and also proved that it can be obtained as the set of all allocations that are associated with Kruskal's (1956) algorithm to construct minimum cost spanning trees.

Knowing that there are several algorithms to construct minimum cost spanning trees, one can consider associating other allocations to them. This is done in Feltkamp, Tijs and Muto (1994a) and (1994b) for Kruskal's algorithm, and an older algorithm described in Borůvka (1926).

References

Aarts H (1992) Marginal allocations in the core of minimal chain games, *working paper no. 1103, Dept. of Applied Math, University of Twente, Enschede, The Netherlands.*

Aarts H and Driessen Th (1993) The irreducible core of a minimum cost spanning tree game, *Zeitschrift für Operations Research* 38: 163–174.

Bird CG (1976) On cost allocation for a spanning tree: A game theoretic approach. *Networks* 6: 335–350.

Boruvka O (1926a) O jistém problému minimálním. *Práce Mor. Prírodoved. Spol. v Brne (Acta Societ. Scient Natur. Moravicae)* 3: 37–58.

Boruvka O (1926b) Príspevek k resení otázky ekonomické stavby elektrovodních sítí. *Elektrotechnický obzor* 15: 153–154.

Claus A and Kleitman DJ (1973) Cost allocation for a spanning tree. *Networks* 3: 289–304.

Dijkstra EW (1959) A note on two problems in connection with graphs, *Numerische Mathematik* 1: 269–271.

Feltkamp V (1998) Strategic minimum cost spanning tree games, *mimeo.*

Feltkamp V, Tijs S, and Muto S (1994a) On the irreducible core of minimum cost spanning extension games, *CenTER DP 94106*, Tilburg University, The Netherlands.

Feltkamp V, Tijs S, and Muto S (1994b) Minimum cost spanning extension problems: the proportional and decentralized rules, *CenTER DP 9496*, Tilburg University, The Netherlands.

Graham RL and Hell P (1985) On the history of the minimum spanning tree problem. *Annals of the History of Computing* 7: 43–57.

Granot D and Huberman G (1981) On minimum cost spanning tree games, *Mathematical programming* 21: 1–18.

Granot D and Huberman G (1982) The relationship between convex games and minimal cost spanning tree games: A case for permutationally convex games, *SIAM J. Alg. and Discr. Meth.* 3: 288–292.

Granot D and Huberman G (1984) On the core and nucleolus of minimum cost spanning tree games, *Mathematical programming* 29: 323–347.

Kuipers J(1993) On the core of information graph games. *Int. J. of Game Th.* 21: 339–350.

Kruskal JB(1956) On the shortest spanning subtree of a graph and the traveling salesman problem, *Proc. American Math. Soc.* 7: 48–50.

Prim RC (1957) Shortest connection networks and some generalizations, *Bell Systems Techn. J.* 36: 1389–1401.

CHAPTER 7

HOW TO SHARE RAILWAYS INFRASTRUCTURE COSTS?*

VITO FRAGNELLI (fragnell@al.unipmn.it)
Università del Piemonte Orientale
Corso Borsalino 54, 15100, Alessandria, Italy

IGNACIO GARCÍA-JURADO (ignacio@zmat.usc.es)
Department of Statistics and OR, Faculty of Mathematics
Universidad de Santiago de Compostela
15771 Santiago de Compostela, Spain

HENK NORDE (h.norde@kub.nl)
Department of Econometrics, Tilburg University
P.O. Box 90153, 5000 LE Tilburg, The Netherlands

FIORAVANTE PATRONE (patrone@dima.unige.it)
Department of Mathematics, University of Genoa
Via Dodecaneso 35, 16146 Genova, Italy

STEF TIJS (s.h.tijs@kub.nl)
Department of Econometrics and CentER, Tilburg University
P.O. Box 90153, 5000 LE Tilburg, The Netherlands

Abstract. In this paper we propose an infrastructure access tariff in a cost allocation problem arising from the reorganization of the railway sector in Europe. To that aim we introduce the class of infrastructure cost games. A game in this class is a sum of airport games and what we call maintenance cost games, and models the infrastructure costs (building and maintenance) produced when a set of different types of trains belonging to several agents makes use of a certain infrastructure. We study some properties of infrastructure cost games and provide a formula for the Shapley value of a game in this class. The access tariff we propose is based on the Shapley value of infrastructure cost games.

* The financial support of the EuROPE-TRIP (European Railways Optimisation Planning Environment-Transportation Railways Integrated Planning) research project is gratefully acknowledged. I. García-Jurado also acknowledges the Spanish Ministerio de Educación y Cultura for financial support through project PB97-0550-C02-02.

1. Introduction

In this paper we deal with a cost allocation problem arising from the reorganization of the railway sector in Europe, after the application of the EEC directive 440/91 and the EC directives 18/95 and 19/95, which involve the separation between infrastructure management and transport operations. In this situation two main economic problems arise. One is to allocate the track capacity among the various operators. This issue has been treated, for instance, in Nilsson (1995), Brewer and Plott (1996) and Bassanini and Nastasi (1997). The second problem is to determine the access tariff that the railway transport operators must pay to the firm in charge of the infrastructure management for a particular journey. This tariff should take into account several aspects such as the a priori profitability and social utility of the journey, congestion issues, the number of passengers and/or goods transported, the services required by the operator, infrastructure costs, etc. The tariff is conceived in an additive way, i.e. as the sum of various tariffs corresponding to the various aspects to be considered.

The main motivation of this paper is a practical one. We were approached by *Ferrovie dello Stato*[1] (the italian national railway company) to study how the infrastructure costs should be allocated to the operators through a fair *infrastructure access tariff* (i.e. we were asked to define one part of the additive access tariff: that corresponding to the infrastructure costs). In this work we treat this problem from a game theoretical point of view, making use of the Shapley value. The Shapley value is a very important solution concept for TU-games, which has excellent properties and has been applied successfully in cost allocation problems (see Shapley (1953), Tijs and Driessen (1986), Young (1994) and Moulin and Shenker (1996)). Moreover, in our particular problem, it is especially appropriate because of the following two reasons.

1. It is well-known that the Shapley value is an additive solution. This feature fits well with the "additive nature" of the access tariff, as commented above.
2. In this paper we will show that the infrastructure access tariff based on the Shapley value can be computed very easily (using, once more, the additivity of the Shapley value). In a practical environment this is certainly an important property. Take into account that a very big amount of fees will have to be computed by the infrastructure manager every new season, so computational issues become highly relevant.

---

[1] Ferrovie dello Stato is the coordinator of the EuROPE-TRIP research project, sponsored by the European Community. Formally, our research has been requested and financed by the European Community.

Let us now describe informally the problem we are facing. Consider a railway path (for instance, Milano-Roma), that is used by different types of trains belonging to several operators, and consider the problem of dividing among these trains the infrastructure costs. Clearly it is a problem of joint cost allocation. To settle the question, one can see the infrastructure as consisting of some kinds of "facilities" (track, signalling system, stations, etc.). Different groups of trains need these facilities at different levels: for example, fast trains need a more sophisticated track and signalling system, compared to local trains, for which instead station services are more important (particularly in small stations).

So, a straightforward approach can be that of viewing the infrastructure as a "sum" of different facilities, each of them required by the trains at a different level of cost.

Furthermore, infrastructure costs can be seen as the sum of "building" costs and "maintenance" costs (for a better understanding of the distinction between these two types of costs, we refer to the example in section 4). If we consider only building costs, especially in the case of a single facility, we are facing a problem similar to the so-called "airport game" (see, for instance, Littlechild and Owen (1973) and Dubey (1982)). For what concerns maintenance costs, it seems to be a reasonable first order approximation to assume that they are proportional both to the building costs and to the number of trains that use the facility.

Similar considerations extend to related problems: for example the costs for a bridge, to be used by small and big cars. There are building costs, that are different in the case of a bridge for small or big cars, and maintenance costs, that can be assumed to be proportional to the number of vehicles using the bridge, and to the kind of bridge needed.

In this paper we analyze these infrastructure cost games (sums for various facilities of a building cost game and a maintenance cost game) from the point of view of the Shapley value. In section 2 we introduce and briefly study the infrastructure cost games. In section 3 we provide a simple expression of the Shapley value for this class of games. In section 4 we elaborate an example where we apply the models and results presented in sections 2 and 3.

2. Infrastructure Cost Games

For simplicity, we concentrate first on infrastructure cost games when we are dealing with the building and maintenance of one facility. To begin with, we recall the definition of an "airport game".

DEFINITION 2.1.   Suppose we are given k groups of players $g_1, \dots, g_k$ with $n_1, \dots, n_k$ players respectively and k non-negative numbers $b_1, \dots, b_k$.

The *airport game* corresponding to $g_1, \ldots, g_k$ and $b_1, \ldots, b_k$ is the cooperative (cost) game $\langle N, c \rangle$ with $N = \cup_{i=1}^{k} g_i$ and cost function c defined by

$$c(S) = b_1 + \cdots + b_{j(S)}$$

for every $S \subseteq N$, where $j(S) = \max\{j : S \cap g_j \neq \emptyset\}$.

Airport games are cost games for the building of one facility (for instance, a landing strip) where the wishes of the coalitions are linearly ordered. Coalitions desiring a more sophisticated facility (a larger landing strip) have to pay at least as much as coalitions desiring a less sophisticated facility (a smaller landing strip). Every b_i represents the extra building cost that should be made in order that a facility that can be used by players in groups $g_1, \ldots g_{i-1}$ can also be used by the more sophisticated players in group g_i. Airport games are known to be concave. Consequently, the Shapley value of such a game provides a core element. Sometimes we will refer to an airport game as a *building cost game*. Denote by $B(g_1, ..., g_k)$ the set of all building cost games with groups of players $g_1, ..., g_k$.

In airport games costs for the building of one facility are modeled. Now we consider the maintenance costs of this facility, which lead to the class of "maintenance cost games". Basic assumptions are that maintenance costs are increasing with the degree of sophistication of the facility and that maintenance costs are proportional to the number of users.

DEFINITION 2.2.　Suppose we are given k groups of players $g_1, \ldots, g_k$ with $n_1, \ldots, n_k$ players respectively and $k(k+1)/2$ non-negative numbers $\{\alpha_{ij}\}_{i,j \in \{1,...,k\}, j \geq i}$. The *maintenance cost game* corresponding to $g_1, \ldots, g_k$ and $\{\alpha_{ij}\}_{i,j \in \{1,...,k\}, j \geq i}$ is the cooperative (cost) game $\langle N, c \rangle$ with $N = \cup_{i=1}^{k} g_i$ and cost function c defined by

$$c(S) = \sum_{i=1}^{j(S)} |S \cap g_i| A_{ij(S)} \tag{1}$$

for every $S \subseteq N$, where $A_{ij} = \alpha_{ii} + ... + \alpha_{ij}$ for all $i, j \in \{1, ..., k\}$ with $j \geq i$.

The interpretation of the numbers α_{ij} and A_{ij} is the following. Suppose that one player in g_i has used the facility. In order to restore the facility up to level i (the level of sophistication desired by this player) the maintenance costs are $A_{ii} = \alpha_{ii}$. If, however, the facility is going to be restored up to level $i+1$, then extra maintenance costs α_{ii+1} will be made. So, in order to restore the facility up to level j (with $j \geq i$) the maintenance costs are $A_{ij} = \alpha_{ii} + \alpha_{ii+1} + ... + \alpha_{ij}$. Hence, $c(S)$ represents the maintenance costs corresponding to the facility up to the level $j(S)$ (so that all the players in

S can use it), after all players in S have used it. Observe that, for every $i \leq j$, the more sophisticated the facility is (the larger j is), the higher the maintenance costs produced by a player in g_i are. In Section 4 we provide an example which illustrates the above definition of a maintenance cost game.

We denote by $M(g_1, ..., g_k)$ the set of all maintenance cost games with groups of players $g_1, ..., g_k$. Obviously, to characterize a game $\langle N, c \rangle \in M(g_1, ..., g_k)$ it is equivalent to give either the set of parameters $\{\alpha_{ij}\}_{i,j \in \{1,...,k\}, j \geq i}$ or the set of parameters $\{A_{ij}\}_{i,j \in \{1,...,k\}, j \geq i}$.

The following decomposition of a maintenance cost game $\langle N, c \rangle \in M(g_1, ..., g_k)$ will be useful. For every $S \subseteq N$,

$$
\begin{aligned}
c(S) &= \sum_{i=1}^{j(S)} |S \cap g_i| A_{ij(S)} = \\
&= \sum_{i=1}^{j(S)} |S \cap g_i| (\alpha_{ii} + ... + \alpha_{ij(S)}) = \sum_{i=1}^{k} \sum_{j=i}^{k} \alpha_{ij} c^{ij}(S),
\end{aligned}
$$

where

$$
c^{ij}(S) = \begin{cases} |S \cap g_i| & \text{if } j \leq j(S) \\ 0 & \text{if } j > j(S) \end{cases}
$$

for all $i, j \in \{1, ..., k\}$ with $j \geq i$.

We know that building cost games are concave. The following result shows that this is not true for maintenance cost games. Moreover, it shows that maintenance cost games are essentially neither concave nor balanced.

THEOREM 2.1. *Let $\langle N, c \rangle$ be the maintenance cost game corresponding to $g_1, \dots, g_k$ and $\{\alpha_{ij}\}_{i,j \in \{1,...,k\}, j \geq i}$. Then the following four statements are equivalent:*

(1) $\langle N, c \rangle$ is concave

(2) $\langle N, c \rangle$ is balanced

(3) $\sum_{i \in N} c(i) \geq c(N)$

(4) $\alpha_{ij} = 0$ for every $j > i$.

Proof. The implications (1) $\Rightarrow$ (2) and (2) $\Rightarrow$ (3) are clear. For the implication (3) $\Rightarrow$ (4) suppose that (3) holds. Then

$$
\sum_{i=1}^{k} \sum_{j=i}^{k} \alpha_{ij} n_i = c(N) \leq \sum_{i \in N} c(i) = \sum_{i=1}^{k} \alpha_{ii} n_i
$$

which implies that $\alpha_{ij} = 0$ for every $j > i$. For the implication (4) $\Rightarrow$ (1) suppose that (4) holds. Note that c^{ii} defined as above is an additive characteristic function for every $i \in \{1, ..., k\}$. Hence, c can be expressed as a non-negative combination of additive characteristic functions. Thus, $\langle N, c \rangle$ is concave. ∎

Now we can introduce the class of infrastructure cost games.

DEFINITION 2.3.   A *one facility infrastructure cost game* with groups of players $g_1, \ldots, g_k$ is the cooperative (cost) game $\langle N, c \rangle$ with $N = \cup_{i=1}^k g_i$ and cost function $c = c_b + c_m$ such that $\langle N, c_b \rangle \in B(g_1, \ldots, g_k)$ and $\langle N, c_m \rangle \in M(g_1, \ldots, g_k)$. An *infrastructure cost game* with groups of players $g_1, \ldots, g_k$ is the cooperative (cost) game $\langle N, c \rangle$ with $N = \cup_{i=1}^k g_i$ and cost function $c = c^1 + \ldots + c^l$ such that, for every $r \in \{1, \ldots, l\}$, $\langle N, c^r \rangle$ is a one facility infrastructure cost game with groups of players $g_{\pi^r(1)}, \ldots, g_{\pi^r(k)}$, where π^r is a permutation of $\{1, \ldots, k\}$.

From the definition above we see that a one facility infrastructure cost game is the sum of a building cost game plus a maintenance cost game with the same groups of players ordered in the same way. An infrastructure cost game is the sum of a finite set of one facility infrastructure cost games with the same groups of players but, perhaps, ordered in a different way. This means that group i can require a higher level of sophistication than group j for facility r, whereas group j requires a higher level of sophistication than group i for facility s. Because of this reason, it is not true that every infrastructure cost game is a one facility infrastructure cost game. An interesting consequence of Theorem 2.1, the concavity of airport games and the additivity of the Shapley value is the following. Since an infrastructure cost game is the sum of building cost games and maintenance cost games, then its Shapley value is the sum of allocations, which are moreover core allocations for those such games having a non-empty core.

The class of infrastructure cost games is the model we designed to solve the practical problem which motivates this work: how to allocate in a fair way the infrastructure costs to the users of a certain railway path. A game in our class describes the infrastructure costs imputable to every possible collection of users. Now we have to choose an allocation rule which allocates the total cost to the users. As we announced in the introduction of this paper, we chose the Shapley value because of the two reasons already discussed. The access tariff we propose for a certain path in a certain time period is simply the Shapley value of the infrastructure cost game corresponding to this path and time period.

Note that an infrastructure cost game is the sum of a finite collection of airport games and maintenance cost games. It is well known that there is a simple expression of the Shapley value for airport games (see Littlechild and Owen, 1973). In the next section we obtain a simple expression of the Shapley value for maintenance cost games. Hence, since the Shapley value is additive, we can compute easily the Shapley value of an infrastructure cost game even when the number of players is large, which will be the case in practice: take into account that the players here are the trains using the path in a certain period. Thus, we are proposing an access tariff system which is at the same time reasonable (based on a general theory of fairness) and computable in an efficient way.

3. The Shapley Value of a Maintenance Cost Game

This section contains a theorem providing a simple expression of the Shapley value of a maintenance cost game.

THEOREM 3.1. *Let* $\langle N, c \rangle$ *be the maintenance cost game corresponding to the groups* $g_1, ..., g_k$ *(with* $n_1, ..., n_k$ *players respectively) and to* $\{\alpha_{lm}\}_{l,m \in \{1,...,k\}, m \geq l}$. *Then, for every* $i \in N$,

$$\varphi_i(c) = \alpha_{j(i)j(i)} + \sum_{m=j(i)+1}^{k} \alpha_{j(i)m} \frac{n_m + ... + n_k}{n_m + ... + n_k + 1}$$

$$+ \sum_{m=2}^{j(i)} \sum_{l=1}^{m-1} \alpha_{lm} \frac{n_l}{(n_m + ... + n_k)(n_m + ... + n_k + 1)},$$

where $\varphi_i(c)$ *denotes the* i-*th component of the Shapley value of the game* $\langle N, c \rangle$ *and* $j(i)$ *is the group to which* i *belongs (i.e.* $i \in g_{j(i)}$).

Proof. Recall that $c = \sum_{l=1}^{k} \sum_{m=l}^{k} \alpha_{lm} c^{lm}$ where

$$c^{lm}(S) = \begin{cases} |S \cap g_l| & \text{if } m \leq j(S) \\ 0 & \text{if } m > j(S). \end{cases}$$

Then, since the Shapley value is linear,

$$\varphi_i(c) = \sum_{l=1}^{k} \sum_{m=l}^{k} \alpha_{lm} \varphi_i(c^{lm})$$

for all $i \in N$. It is clear that, for every $l \in \{1, ..., k\}$, c^{ll} is an additive characteristic function and that

$$\varphi_i(c^{ll}) = \begin{cases} 1 & \text{if } i \in g_l \\ 0 & \text{in any other case.} \end{cases} \tag{2}$$

Suppose now that $l < m$. In this case only players in $g_l \cup (\cup_{r=m}^{k} g_r)$ are not null players. By symmetry we may put $\varphi_i(c^{lm}) = a$ for every $i \in g_l$ and

$\varphi_i(c^{lm}) = b$ for every $i \in \cup_{r=m}^{k} g_r$. In order to compute a take $i \in g_l$ and note that for every $S \subseteq N\backslash\{i\}$ we have

$$c^{lm}(S \cup \{i\}) - c^{lm}(S) = \begin{cases} 0 & \text{if } j(S) < m \\ 1 & \text{else.} \end{cases}$$

So, if the players of N are ordered at random, a is the probability that player i has at least one predecessor in $\cup_{r=m}^{k} g_r$. Equivalently, if the players of N are ordered at random, a is the probability that player i is not the first player of the players in $\{i\} \cup (\cup_{r=m}^{k} g_r)$. Consequently,

$$a = \frac{n_m + ... + n_k}{n_m + ... + n_k + 1}. \tag{3}$$

Thus, by symmetry and efficiency,

$$b = \frac{n_l - n_l a}{n_m + ... + n_k} = \frac{n_l}{(n_m + ... + n_k)(n_m + ... + n_k + 1)}. \tag{4}$$

Now, in view of (2), (3) and (4) the proof is concluded. ∎

As we mentioned above, the Shapley value of the corresponding infrastructure game is our proposal to share railways infrastructure costs. It is clear that, using Theorem 3.1 and the formula for the Shapley value of an airport game, the computations that should be made are not difficult; however, the potentially very large amount of data that will have to be handled to compute a very large collection of fees makes necessary to have a good computer program to do it. For this purpose, we have prepared a software package that will be delivered to *Ferrovie dello Stato*, the coordinator of EuROPE-TRIP. The name of this package is ShRInC (Sharing Railways Infrastructure Costs). It has been created with the collaboration of Luisa Carpente and Claudia Viale.

Obviously, from a game theoretical point of view, there are many interesting questions concerning infrastructure cost games that have not been treated here. The main motivation of this paper is to report the practical solution we proposed for the real problem of allocating railways infrastructure costs. In Norde et al (1999), we study other game theoretical properties of infrastructure cost games.

4. An Example

In this section we illustrate our solution with an example. We shall elaborate it on data taken from Baumgartner (1997). The aim of that paper is to provide "order of magnitude" of costs concerning the railway system:

we shall exploit it to analyze a rough but realistic example. In practical models, making a realistic example uses to be an enlightening exercise. Here, for instance, the example we are proposing shows that our building or maintenance cost games do not necessarily correspond to real building or maintenance costs. Actually, the costs for one facility can be decomposed into:

— a fixed part (in the sense that it does not depend on the number of players), that corresponds to the building cost game associated with this facility, and

— a variable part (in the sense that it is proportional to the number of players), that corresponds to the maintenance cost game part.

For simplicity, we shall concentrate on a single element (the track), even if Baumgartner (1997) provides data also for other elements (line, catenary, signalling and security system, etc.), that can be analyzed in a similar fashion. If we consider one kilometer of track, from Baumgartner (1997) we get two kind of costs[2], that depend on the type of train (slow/fast) and on the number of trains running. More precisely, we have both *renewal costs* and *repairing costs*. According to this division of costs we will divide the track into two facilities: "track renewal" and "track repairing".

Renewal costs can be approximated by the following formula:

$$RWC = 0.001125X + 11,250$$

where RWC are the renewal costs per kilometer and per year (expressed in swiss francs) and X measures the "number" of trains, expressed in yearly TGCK (TGCK means Tons Gross and Complete per Kilometer).

So, if we assume for ease of exposition that all of the trains running are of the same weight, the facility "track renewal" has a fixed component (to be included in our building costs), and a part which is proportional to the number of trains running (to be included in our maintenance costs). If the assumption of equal weight cannot be sustained, our model still fits: simply divide trains into groups of similar weight. In such a case each group will have different unitary maintenance costs.

Similarly, for the facility "track repairing", costs can be given by analogous formulas:

$$RPC_s = 0.001X + 10,000$$

$$RPC_f = 0.00125X + 12,500.$$

RPC_s denotes the repairing costs (in swiss francs) per kilometer and per year of a track prepared only for slow trains, whereas RPC_f denotes the

_____

[2] We assumed the weight of 50Kg for a meter of rail and made a linear approximation of the costs given in table 2 of Baumgartner (1997).

repairing costs (in swiss francs) per kilometer and per year of a track prepared for all trains. X denotes the same as before.

So, consider one kilometer of line, which will be used this year by a total weight of 10^7 TGCK (corresponding to 20,000 trains, assuming a weight per train of approximately 500 tons). Assume that 5,000 trains are fast and that the remaining are slow. The infrastructure cost game that can be used to allocate the costs is $\langle N, c \rangle$ given by:

- $N = g_1 \cup g_2$, g_1 being the set of slow trains ($n_1 = 15,000$) and g_2 being the set of fast trains ($n_2 = 5,000$).

- $c = c^1 + c^2$, c^1 and c^2 being one facility infrastructure cost games both having the same groups of players and ordered in the same way: g_1, g_2.

Now, c^1 and c^2 are characterized by the following parameters.

- $c^1 : b_1^1 = 11,250;\ b_2^1 = 0;\ \alpha_{11}^1 = 0.5625;\ \alpha_{12}^1 = 0;\ \alpha_{22}^1 = 0.5625.$
- $c^2 : b_1^2 = 10,000;\ b_2^2 = 2,500;\ \alpha_{11}^2 = 0.5;\ \alpha_{12}^2 = 0.125;\ \alpha_{22}^2 = 0.625.$

Hence, making use of Theorem 3.1 and the formula for the Shapley value of an airport game, it is easy to check that, if $\varphi_s(c)$ and $\varphi_f(c)$ denote the Shapley value of a slow and a fast train respectively, then:

- $\varphi_s(c) = \frac{b_1^1}{n_1+n_2} + \alpha_{11}^1 + \frac{b_1^2}{n_1+n_2} + \alpha_{11}^2 + \alpha_{12}^2 \frac{n_2}{n_2+1} = 2.25$

- $\varphi_f(c) = \frac{b_1^1}{n_1+n_2} + \alpha_{22}^1 + \frac{b_1^2}{n_1+n_2} + \frac{b_2^2}{n_2} + \alpha_{22}^2 + \alpha_{12}^2 \frac{n_1}{n_2(n_2+1)} = 2.75.$

These are the fees, in swiss francs, that every slow and fast train (respectively) should pay per kilometer of track used, according to our solution. Clearly, in front of a specific allocation problem regarding a specific line, with specific transport operators and trains, appropriate data should be collected. Here, we only presented an illustrative approximation to a real example.

References

Bassanini A, Nastasi A (1997) A Market Based Model for Railroad Capacity Allocation. Research Report RR-97.08, Dipartimento di Informatica, Sistemi e Produzione. Università degli Studi di Roma Tor Vergata.

Baumgartner JP (1997) Ordine di Grandezza di Alcuni Costi nelle Ferrovie. Ingegneria Ferroviaria 7:459-469.

Brewer PJ, Plott CR (1996) A Binary Conflict Ascending Price (BICAP) Mechanism for the Decentralized Allocation of the Right to Use Railroad Tracks. International Journal of Industrial Organization 14:857-886.

Dubey P (1982) The Shapley Value as Aircraft Landing Fees Revisited. Management Science 28:869-874.

Littlechild S, Owen G (1973) A Simple Expression for the Shapley Value in a Special Case. Management Science 20:370-372.

Moulin H, Shenker S (1996) Strategyproof Sharing of Submodular Access Costs: Budget Balance versus Efficiency. Mimeo.

Nilsson JE (1995) Allocation of Track Capacity. CTS Working Paper 1995:1, Centre for Research in Transportation and Society, Dalarna University College, Sweden.

Norde H, Fragnelli V, García-Jurado I, Patrone F, Tijs S (1999) Balancedness of Infrastructure Cost Games. Preprint DIMA 369. Università di Genova.

Shapley LS (1953) A Value for n-Person Games. In: Contributions to the Theory of Games II. H Kuhn and AW Tucker (eds), pp 307-317. Princeton University Press.

Tijs S, Driessen T (1986) Game Theory and Cost Allocation Problems. Management Science 32:1015-1028.

Young P (1994) Cost Allocation. In: Handbook of Game Theory (vol 2). RJ Aumann and S Hart (eds), pp 1193-1235. North-Holland.

CHAPTER 8

WHY PUNISH? NORMS AND REVENGE IN AN EXPERIMENTAL GAME *

URI GNEEZY (gneezy@econ.haifa.ac.il)
Department of Economics
University of Haifa
Haifa, Israel

AVRAHAM STOLER
University of Tel Aviv
Tel Aviv, Israel

Abstract. People punish others for "unfair behavior" even when it is costly for them to do so. In the experiment, the effects of revenge and education were isolated as motives for punishment. Significantly less punishment was observed when either of the two motives was eliminated.

1. Introduction

It is commonly assumed in economics that people act as if they are maximizing their monetary rewards. The cumulative observations of real-life as well as experimental studies suggest, however, that this assumption does not describe the way people actually behave in some situations. One example is the willingness to punish what is perceived as unfair behavior.
Empirical results from games, such as the ultimatum game, continually show that people are willing to lose money in some situations just in order to punish a counterpart for that person's behavior. [See the original investigation by Güth, Schmittberger, and Schwarze (1982), and the surveys in Camerer and Thaler (1995), Güth (1995), and Roth (1995) for the robustness of this phenomenon.]

  This paper reports on an initial experimental investigation of the importance of two motives for punishment. The first is *education* — to educate

---

* We thank Chaim Fershtman for many helpful comments and suggestions.

someone to adhere to specific norms. Norms are relevant in economics. For example, Camerer and Thaler (1995, p.218) argue that "people have simply adopted rules of behavior they think apply to themselves and others, regardless of the situation. They leave tips in restaurants that they never expect to visit again, not because they believe this is really a repeated game, but because it would be rude to do otherwise." On the importance of norms, see also Hoffman, McCabe, and Smith (1996), and Axelrod (1984).

Thus, people may punish others in order to educate them to maintain the norms of "good citizenship." This can mean, for example, punishing "unfair" behavior in order to educate the unfair person to be a "good citizen" and to care about fairness. Even when punishment decreases material payoff, it may be in society's interest. This mechanism is, in fact, the internalization of a positive externality, and itself can be explained as a kind of metanorm (punish all those who are not behaving according to the norm). See Axelrod (1984, p.1100).

The second motive investigated is *revenge*, defined here as suffering losses just for the pleasure of harming an unfair person. Revenge-type behavior is well known in real life ("sweet revenge"). In the economics literature, it has been addressed, for example, by Bolton and Zwick (1993, p.100), who argue that "a player's preference for more money is modified by a preference for disagreement over amounts he perceives as small *relative* to his playing partner's share." It is also argued that people are averse to others' behaving unfairly toward them [see Bolton (1991) and Güth and Tietz (1988)].

The effects of educational and revenge motives in a baseline "punishment game" were compared, first with a game in which the educational motive was eliminated and then with a game in which the possibility of direct revenge was eliminated[1].

2. Experimental design and procedures

The experiment was conducted at the University of Haifa, in November 1997. All participants were first-year undergraduate students in the Economics Department. Each participant played only once and only one of the roles. All participants were paid, privately and in cash, at the end of the experiment.

In the first stage of the experiment, each participant (called an "allocator") was asked to split NIS 30[2] with another participant from a different class, with whom the allocator was randomly matched. Allocators were told

---

[1] See also the recent study by Abbink, Sadrieh, and Zamir (1997), who ask questions similar to ours, but a different experimental setup.

[2] At the time of the experiment, NIS 3.5=$1.

that after they made their choice they would be paid, and that this was the only choice they had to make in the experiment.

The second stage of the experiment employed three experimental treatments. In each treatment participants were told about the procedure for the first stage of the experiment (which included seeing the allocators original instructions). Every participant was then told whether the amount offered by the allocator was (a) more than NIS 20, (b) between NIS 10 and NIS 20, or (c) less than NIS 10.

Treatment 1: Each participant was asked to choose one of two options. One was to divide NIS 40 evenly with the allocator she was matched with. The alternative was to take NIS 12 and leave nothing to her counterpart. Participants in this treatment were told that their counterparts would be informed of the choice they had to make and of their decision. It was also made clear that this was the last stage of the experiment and that no more decisions would have be made.

Treatment 2: Participants were asked to make the same decision as in treatment 1. The difference was that participants were told that the allocators would not know about the decision that had to be made. Instead, some of the allocators would merely be paid a "bonus" for no apparent reason. As in treatment 1, participants were assured that this was the last part of the experiment.

Treatment 3: Participants were given information about the choice made by a "third allocator"; that is, not the person who had to decide how much money to give them. (Participants were not told at this point how much money was to be shared with them by "their" allocator.) They were then asked to decide between two options similar to those in treatment 1. The difference was that the division was to be made between the participant and the "third player," not the allocator who decided how much money to give him.

3. Results and discussion: why punish?

Why do people punish others for unfair behavior? Two motives were mentioned in the introduction: education and revenge. In the context of our experiment, we may say that the responder has an *educational* motive when her utility increases from signaling to the allocator that there are some norms that should be adhered to in order to avoid punishment. We say that the responder has a *revenge* motive when her utility increases from reducing the utility of the allocator. The experiment reported in this paper was designed to supply some empirical evidence regarding these motives[3].

---

[3] The literature arises other motives for punishment. For example, in a repeated game, a behavior that is costly in the present may be optimal, because it has long-term effects on

In the basic game (treatment 1), 16 of 41 (39%) participants who received offers of less than NIS 10 chose to punish their unfair counterpart. This confirms what has been observed in many other experimental games; namely, that some people are willing to suffer monetary losses in order to punish what they conceive of as unfair behavior.

In treatment 2 the level of information that the punished player receives was controlled, thereby eliminating the possibility of educating. That is, when the allocators do not know (not even ex post) that the payoff was influenced by the behavior of another person, they cannot be "educated." This manipulation of information turned out to have a significant effect: only 4 of 40 participants (10%) who received an offer of NIS 10 or less chose to punish ($z = 3.0286$, $p < .01$). Accordingly, we reject the hypothesis that "educational" motives do not influence the decision to punish. Apparently, a reduction in the monetary reward given to the unfair player is not the sole purpose of punishment. It is also important to the punishing player that the unfair player knows that she was punished.

In treatment 3, participants could not punish those who had been unfair toward them. They could, though, punish another participant who had been unfair toward somebody else. We found that only 6 of 40 participants (15%) who could punish an allocator who had made an offer of NIS 10 or less chose to punish. (This result is also significant with $z = 2.4304$, $p < .01$). The educational aspect of punishment is not affected (that is, the willingness to punish in order to educate should not be affected by the identity of the player who suffered the unfair behavior). But is revenge still as "sweet" as in the original case? Apparently not.

Finally, the difference between the number of participants who punished in treatments 2 and 3 (4 of 40 and 6 of 40 respectively) is not significant ($z = .6765$, $p > .1$)

4. Conclusions

Both education and revenge matters; excluding one of these motives was enough almost to eliminate punishment in the game tested. These findings could be used as building blocks for a descriptive theory that accounts for punishment behavior. In particular, such a theory could help in predicting the circumstances in which we should expect people to punish.

---

the behavior of others; i.e., the players lose something by punishing their opponents in the present, but this punishment will induce more cooperative behavior from the opponents in the future. [See, for example, Osborne and Rubinstein (1994) and Fudenberg and Tirole, (1995)].

References

Abbink, K., A. Sadrieh, and S. Zamir (1997): *The Covered Response Ultimatum Game*, SFB Discussion Paper B-416, University of Bonn.

Bolton, G. (1991): *A Comparative Model of Bargaining: Thoery and Evidence*, American Economic Review, 81:5, 1096-1136.

Bolton, G. and R. Zwick, (1993): *Anonymity Versus Punishment in Ultimatum Bargaining*, Games and Economic Behavior, 10, 95-121.

Camerer, C. and R. Thaler, (1995): *Ultimatums, Dictators and Manners*, Journal of Economics Perspectives, 9:2, 209-219.

Güth, W. (1995): *On Ultimatum Bargaining Experiments: A Personal Review*, Journal of Economic Behavior and Organization, 27:3, 329-44.

Güth, W., R. Schmittberger and B. Schwarze (1982): *An Experimental Analysis of Ultimatum Bargaining*, Journal of Economic Behavior and Organization, 3, 367-388.

Güth, W., and R. Tietz (1988): 'Ultimatum Bargaining for a Shrinking Cake-An Experimental Analysis', in W. Albers, R. Selten, and R. Tietz (eds.): *Bounded Rational Behavior in Experimental Games and Markets*, Lecture Notes in Economics and Mathematical Systems, Vol.314, Berlin, 111-128.

Hoffman, E., McCabe, K., and V. Smith (1996): *Social Distance and Other-Regarding Behavior in Dictator Games*, American Economic Review, 86:3, 653-660.

Roth, A. (1995): 'Bargaining Experiments', In J. Kagel and A. Roth (eds), *The Handbook of Experimental Economics*, Princeton: Princeton University Press.

CHAPTER 9

A GAME-THEORETICAL PERSPECTIVE FOR THE
DETECTION OF TACIT COLLUSION

MICHELE GRILLO (michele.grillo@agcm.it)
Autorità Garante della Concorrenza e del Mercato
Via Liguria 26, Rome (Italy)

Abstract. This paper deals with two conceptually distinct. but connected, issues on the contribution of game-theoretical analysis to the detection of tacitly colluding behaviour in oligopoly. First, the need is emphasised to clearly distinguish between the notion of co-ordination and that of co-operation. Such a distinction puts a relevant constraint on the standard legal approach to collusion which misleadingly emphasises the former notion. Next, a critical review of the now growing literature on semicollusion suggests that sharper inference on collusive conduct should be drawn on the observation of the overall set of the firm's long- and short-run decision variables.

1. Introducing the question: the legal and the economic approach to the detection of collusive behaviour

Nowadays game-theoretical analysis pervasively underlies every relevant issue on competition policy. The economic markets with which the Competition Authorities usually deal are normally oligopolistic markets. Even when a single firm in a monopoly is under scrutiny, what really matters for competition policy is its behaviour towards potential competitors, so the appropriate analytical context is definitely strategic. Thus game theory is equally important in the analysis of mergers, of both horizontal and vertical agreements, of abuse of dominant position, to refer to the threefold partition which is now standard in the European Competition Law and in the Competition Law of every European country.

Notwithstanding this general coverage of game-theoretical analysis for competition policy, I have chosen to focus on one single specific issue, namely tacit collusion. This is of course because of the very sake of sharpness, but also because implicit co-ordination in oligopoly, leading to collusive equilibria, is to my view one of the most challenging questions that competition policy in the European countries faces today, both at the theoretical and at the applied level.

Detecting collusive behaviour is of great relevance to competition policy makers. As in a monopoly, also under collusion, industry profits are maximised at the expense of society's overall surplus. It is by now widely recognised that the loss for society greatly exceeds the standard textbook deadweight loss. As goods and services are priced significantly higher than their production cost, the social division of labour is adversely affected: it is possible for inefficient producers to still remain in the market, by dissipating the collusive rents through their inefficiencies[1]. The conventional antitrust approach to collusion is that collusive behaviour is the result of *co-ordination* among oligopolistic firms, which agree on implementing monopolistic price strategies or on setting quotas. Thus, in order to detect collusion, the antitrust analysis looks at firms' co-ordination, namely at "agreements" and "concerted practices".

The conventional approach to collusion is to my view unsatisfactory. On the one hand, the hopes of detecting collusion, because evidence is found of an explicit collusive agreement on the price - as might more easily have happened in early days when competition statutes were being introduced, mainly in countries with a historical experience of legal cartels - are now rapidly fading away. Today hardly, and only in minor cases, such direct evidence is found[2]. Since oligopolistic firms now take great care not to leave evidence around of a collusive agreement, however they have reached it, the big legal question is raised as to the extent to which one can, from circumstantial evidence, infer the existence of a tacit, or explicit, illegal agreement.

On the other hand, there is much more to the point than the mere fact that a handful of wrong-doers succeed in hiding the evidence of their malfeasance; for, in a precise game-theoretical perspective, explicit communication among social agents is not to be viewed either as a *necessary* nor as a *sufficient* condition for co-operation to arise.

Game-theoretical analysis of rational behaviour in social situations compels us to clearly distinguish between *co-ordination* and *co-operation*. The

---

[1] Collusion likely arises in industries with a small number of producers. Usually non-convexities in production technology (such as economies of scale) are at work, which make the market structurally non-competitive. Under these circumstances, to change the market structure by increasing the number of producers, in order to reduce the likelihood of collusion, would not prove to be an efficient remedy.

[2] Taking the Italian experience in 1997 as an instance, in only one case evidence of price-fixing agreements, in the form of correspondence and meeting minutes, was found (that was in the *Ready-mixed concrete sector*), while in the two most debated cases (i.e., *Glass for food industry* and *CDs, cassettes and vinyl records*) the Italian Competition Authority condemned implicit collusive behaviour.

need for *co-ordination* straightforwardly arises from the intrinsically strategic oligopolistic context. In a strategic context, there is no way for a rational
agent - that is, for an agent that intends to act rationally - to remove or
ignore the inherent interdependence that links its decision problem with the
decision problems of all other "players in the game". The game-theoretical,
Nash equilibrium concept in social situations is thoroughly embedded in
the theory of social action, which according to Max Weber is to be defined
as "action which takes account of the behaviour of others and is thereby
orientated in its course"[3]. No "equilibrium" in a social context can be
reached unless players are able to understand and solve the problem of
co-ordinating on a common behavioural rule.

Conceptionally and operationally, *co-operation* is to be kept distinct
from *co-ordination*. Social agents *co-ordinate* in order that each one's behaviour be rationally "orientated in its course". They *co-operate* when they
select a specific common course of actions which lead to a particular social
equilibrium whereby the surplus that accrue to them is maximised. Surely,
collusion - that is the *co-operative* oligopolistic equilibrium, whereby industry profits are maximised at the exprense of consumers' and society's
overall surplus - can hardly ensue, absent co-ordination. But, in the light
of the co-ordination required to attain it, collusion need not per se be
distinguishable from any other, less co-operative, equilibrium. Nor, in a
different perspective, is the nature and quality of co-ordination generally
connected with the degree of collusion.

On the one extreme, as game-theoretical analysis of tacit co-operation
has made it clear, explicit communication is not a necessary condition for
co-operative, that is, collusive, behaviour to arise. Indeed, the need for
explicit co-ordination on a co-operative equilibrium solution can be min-
imised when collusion simply is the result of conventional adherence to well
established social norms of conduct - as it surely can be under a number of
plausible circumstances. On the other extreme, there can be circumstances
where the attainment of even a non co-operative equilibrium requires a
substantial amount of explicit co-ordination among oligopolistic firms. The
practical need for explicit co-ordination even in a non-cooperative equilibrium can easily be understood whenever the industry context admits of
multiple equilibria. However, the logical need for co-ordination arises from
the very concept of rational behaviour in social equilibrium situations: as
the rationality of choice inherently depends on each player's expectations
about the behaviour of other players, in order to be in equilibrium rational
agents must co-ordinate on a common, rationally-based, expected profile of
choices. My conclusion is straightforward: as in every social context, also

---

[3] See Weber (1922), p.5.

in the oligopolistic context, co-ordination has much more to do with the way in which an equilibrium is attained, rather than with the collusive properties of it.

Here it is where the legal and the economic approaches to collusion suffer to my view from serious, reciprocal, misunderstanding. For, on the one hand, the legal approach, be collusion explicit or tacit, focuses on co-ordination, rather than on co-operation: it is firmly rooted in the search of an instance of the oligopolists' conscious commitment to a common scheme of conduct. In addition, legal scholars want that evidence of this common understanding be distinct and separate from the collusive conduct itself. On the other hand game-theoretically oriented economists perceive that, while a commitment to a common scheme of conduct may sometime be needed to solve a pure co-ordination problem, there can be circumstances in which co-ordination on a co-operative, i.e. collusive, scheme of conduct does not necessarily require a conscious and separate commitment to it, but only adherence to well established social norms of conduct. In other words, the legal approach focuses on the process through which the oligopolists' commitment to a common scheme of conduct is attained, whereas economic analysis focuses on outcomes: it always has to ask what competitive or collusive outcomes may be produced by certain conduct[4].

In a historical perspective, a first step at overcoming the difference between the legal and the economic approach was made by Donald Turner, some forty years ago. In his now-classic contribution[5], Turner recognised the need to go beyond the boundaries of an explicit, verbally communicated, assent to a common course of action for the detection of a collusive agreement, and proposed to simply define agreement for purposes of the Sherman Act in terms of "interdependence of decisions". Turner was suggesting that, rather than focus on whether an agreement could be discerned, the Department of Justice and the Courts had a deeper look at the anticompetitive effects associated with a certain conduct.

Moreover, Turner was also concerned with the possibility of devising an effective remedy to collusive conduct that would not involve the Courts in a regulatory activity. With these purposes in mind, Turner was looking for so-called "facilitating practices", that is, patterns of conduct that could help co-operation among oligopolistic firms and that could easily be remediable by Courts.

The main flaw of Turner's proposal comes from his analytical premise, according to which the fundamental dividing line between competitive and collusive conduct should be interpreted as the one between individual, independent conduct versus joint, interdependent conduct. In his time,

---

[4] See, on the point, Yao and De Santi (1993).
[5] See Turner (1962).

Turner was undoubtedly entitled to ignore the subtleties of strategic reasoning. My challenge here is that Turner's analytical premise today is still shared by the legal approach. In fact, the oligopolistic setting is unfit for the notion of "independent action". Such a notion surely arose from the economic analysis of the price-taking firm in purely competitive markets and was therefrom easily turned into a well-definite legal notion. However, it contrasts with the game-theoretical analysis of social situations which sees co-ordination as an irreducible element of the process itself whereby a social equilibrium is attained. In conclusion, while firms' actions necessarily are interdependent in oligopoly, one can find both conscious co-ordination that leads to a non-collusive equilibrium, and focal, collusive, equilibria supported and co-ordinated by conventional adherence to social norms of conduct.

I do not intend to claim for novelty. Introducing the Antitrust Bulletin's Symposium on tacit collusion in Spring 1993, Jonathan Baker had already asked whether the legal standards for inferring a horizontal agreement from circumstantial evidence properly reflect the contemporary economic understanding of co-ordinated behaviour[6]. Thus, in what follows, I will try to contribute to Baker's question, by expanding on two issues. Firstly, I will review some standard tools implemented by both the European and the Italian Competition Authorities for the detection of collusion. I will try to make it clear that the legal approach puts a relevant, albeit somehow misleading, constraint on the difficult task of distinguishing truly collusive behaviour from other non-collusive oligopolistic conduct. Next I will tentatively turn to economic, i.e., game-theoretical, analysis, to ask whether sharper inference on collusive conduct can be drawn, by bringing under observation an enlarged set of the firms' decision variables. In this perspective, my point will be that the now growing theoretical literature on semicollusion can suggest important insights to this aim. By way of course, clear-cut and robust results are strongly needed in order to attain the adequate standard for a legal proof. However, even short of definite general results, which are not to my knowledge at hand at the present date, Competition Authorities would greatly improve their job, provided they always corroborate the inference of collusive behaviour - from circumstantial evidence of prices or exchange of information - with a thoroughly assessment of the overall set of the oligopolistic firms' strategic decisions.

---

[6] See Baker (1993).

2. The institutional setting: the detection of collusion according to the European Competition Law

In the European Competition Law collusion in oligopoly is catched under Article 85 of the Treaty of Rome. Provided that there exists direct evidence of an agreed, explicit co-ordination on the price, a sort of rule per se applies. This may contrast, in principle, with the game-theoretical result that explicit co-ordination per se need not be evidence of co-operative, that is, collusive conduct, thus suggesting that further *substantial* analysis of the collusive outcome attained should in any event be in order.

However, I will not dwell on this point, since I prefer to tackle the same basic theoretical question from the opposite perspective of how to detect collusion when such a direct evidence is lacking. Under these circumstances, Competition Authorities resort to evidence of parallel behaviour - usually on price - trying to interpret it as the result of a "concerted practice".

In 1972, in the *Dyestuffs* case, the European Court of Justice defined a "concerted practice" as: "...a form of co-ordination between undertakings which, without having been taken to the stage where an agreement properly so-called has been concluded, knowingly substitutes practical co-operation between them for the risks of competition"[7]. However, the Court refused to identify parallelism of behaviour with a concerted practice, suggesting that, although evidence of parallel behaviour can be taken as a serious hint of "concertation", there may be in oligopoly forms of parallelism, such as for an instance a barometric price leadership, which are beyond the reach of Article 85. Hence, in absence of an explicit agreement, it is essential to explicitly rule out any possible alternative explanation of parallel behaviour other than collusion[8].

By recognising that parallel behaviour is a necessary but not a sufficient condition for a concerted practice, the Court was undoubtedly right. In a sense, the views expressed by the Court just amount to saying that co-

---

[7] Cases 48-57/89 ICI v. Commission (1972) at para 64.

[8] In a sense, the European Court of Justice's approach logically complements the so-called "but-for" test in the US Competition Law, which has been more recently developed and is explicitly based on the application of standard game-theoretical analysis of tacit collusive equilibria. Whereas the European approach requires that any other explanation of parallel behaviour different from collusion be explicitly ruled out, in the US, Courts cannot infer collusive conduct from parallel behaviour unless the following "but-for" test is successfully undergone: (i) a short-run incentive to deviate from parallel behaviour exists; (ii) adherence to the parallel behaviour is not a dominant strategy; (iii) evidence of a recourse to a sort of trigger strategy in the long-run is at hand; (iv) interdependence that depends on downstream or across-time reciprocity can accurately be screened out from interdependence that does not; (v) no direct customer benefit is evident; (vi) the observed parallel behaviour is "remediable" in the sense of Turner. See Yao and De Santi (1993), pp.123ff.

ordinated behaviour is but an element of the social equilibrium, which has
not in itself anything to do with the collusive properties of it. However, when
insisting on the essentiality of explicitly ruling out any possible alternative
explanation of parallel behaviour other than collusion, the Court was still
implicitly looking for evidence of a conscious commitment by the parties to
a common scheme of conduct.

   This was made explicit twenty years later, in the Wood Pulp case. In
1984 the Commission had gone a long way towards finding a concerted
practice in the Wood Pulp case on economic evidence alone, arguing that
the market structure could not explain the observed parallel behaviour but
for the result of a collusive understanding. However, in 1993, on appeal, the
Court of Justice held a different view of the market structure, according to
which non-collusive reasons for the parallelism of prices could be devised[9].

   What is striking in the conclusion of the Wood Pulp case are the implications of the fact that the Court laid on the Commission the daunting
burden of proving that no other possible explanation different from concertation exists for parallel behaviour of prices. Indeed, given the logical
impossibility of the task of excluding each from a potentially unlimited
number of possible explanations, the practical implication of the Court's
decision was the need to resort back to evidence of a concerted practice in
order to infer collusion from observation of parallelism of prices. Here again,
in the same case, "concerted practices" were defined by Advocate General
Darmon as referring to "reciprocal communications between competitors
with the aim of giving each other assurances as to their conduct on the
market"[10], that is in terms of verbal co-ordination. Indeed, since the Wood
Pulp case a simple way to prove concertation is to look for evidence of
exchange of information.

   The Italian experience is akin to the European story. To refer to the two
most debated, recent, cases, in the *Glass for food industry* case, evidence
was found of a parallel behaviour in wholesale prices, which was interpreted
as the result of a concerted practice; in *CDs, cassettes and vinyl records*,
evidence was found of concertation on the prices charged to retailers. Both
cases concerned oligopolistic markets characterised by high entry barriers
and wide product differentiation.

   In both cases the Italian Competition Authority endeavoured to rule
out any possible alternative explanation of parallel behaviour other than
collusion. In the first case, the parallel behaviour was interpreted as the
result of a concerted practice consisting in a periodical exchanging of information about competitors' production, sales and market shares. In the

---

[9] cases C-89/85 A. Ahlstrom Oy and others v. Commission (1993).

[10] cases C-89/85 A. Ahlstrom Oy and others v. Commission, opinion of the Advocate
General Darmon delivered on 7 July 1992.

Record industry case, the same conclusion was based on evidence of regular exchange of information, again favoured by the intense co-ordination of the parties' commercial behaviour in a lot of promotional activities.

Before turning to economic analysis I will briefly refer to the suggestion, put forth by some legal scholars[11], according to which (tacit) collusion can be tackled by directly pursuing the effects of an implicit agreement. This can be attained by condemning abnormal, anticompetitive, prices under Article 86 of the Treaty of Rome, as an "abuse of collective dominance", thus avoiding the need to prove the existence of a concerted practice. For the sake of brevity I will not expand on this second direction of analysis. However, I can hardly conceal my personal view, that substituting the concept of "abuse of collective dominance" for the concept of "agreement" in order to pursue (tacit) collusion, is highly unsatisfactory if not thoroughly misleading. The need for abandoning the way of looking at collusion under Article 85, that is as an agreement of some sort, because of the very difficulty of proving the existence of the agreement itself, then resorting to a concept of abusive joint dominance, is to my view evidence of two weaknesses of the legal approach: (i) on the one hand, the result is ignored that, from game-theoretical analysis, even evidence of explicit co-ordination may not be sufficient to prove collusion, since co-ordination is an element of the process of attaining an equilibrium per se, independently of the collusive properties of it; (ii) on the other hand, the opportunity is refused to catch under Article 85 of the Treaty of Rome relevant evidence of substantial collusive behaviour that economic analysis could explain, even in absence of an explicit and separate evidence of the common understanding to co-operate.

3. Enlarging the set of the firms' decision variables: what can we therefrom infer for collusive behaviour?

I now turn to economic analysis for helpful predictions about the overall behaviour of colluding firms. By so doing, I will wonder whether collusion can be correctly inferred by properly separating the ambiguous issue of co-ordination from the substantial issue of co-operation. Let me first make it clear that I am not looking for predictions that directly involve prices, or even profits. Surely, these variables would be the pre-eminent indicators of collusive behaviour, provided that adequate information on *production* costs be at hand. However, it is a fundamental tenet of economic analysis that information on *production*, that is *minimum*, costs can only be obtained *through the market*: as von Hayek puts it, were this information obtainable independently, the theoretical justification for preferring the market, as a

---

[11] See, for instance, Whish and Sufrin (1993).

superior form of social organisation of exchange and production, over centralisation and planning would be lost, and competition policy would turn itself into a regulatory policy. I will instead look for adequate inference on collusion from observation of the whole set of the firms' decision variables: oligopolistic firms are indeed concerned not only with short-run decisions on prices and quantities but also with several, less reversible, long-run, decisions, such as on product differentiation, capacity, advertising, R&D, and so on[12].

It is common, in modern literature on Industrial Organisation[13], to assume that collusion normally is much more difficult to obtain on the long-run, than on the short-run, variables. The intuition behind this assumption is that, while firms have the strongest reasons to escape, as far as they can, a mutually harming competition on prices, competition on non-price, long-run, decision variables remains the typical weapon in the oligopolistic arena. Its purpose is indeed to structurally improve each firm's position *vis à vis* its own competitors. If a firm succeeds in this purpose, the way in which extra-profits are distributed in the short-run collusive equilibrium are permanently modified. Eventually, provided relative positions have significantly changed, weaker firms can be forced to leave the market, may be through limited and well-oriented bursts of price competition set in motion by stronger rivals.

Due to their different degree of reversibility, short-run and long-run decisions can formally be linked in a multistage decision game: long-run decisions, competitively made in the first stage, are intertwined with short-run decisions on prices and quantities in the second stage. Under this setting firms' incentives to select the long-run strategic variables may be different, depending on whether firms expect that a competitive, or rather a collusive, solution will ensue in the short-run. It is common to label as "semicollusion" the situation in which long-run decisions are competitively made but with the understanding that market collusion will follow in the second stage of the game. On the condition that equilibrium decisions on the long-run variables be predictably different according to whether collusion or competition prevails in the short-run, theoretical models of semicollusion may provide antitrust analysis with useful insights for the detection of collusion.

Before reviewing the main results of this literature, let me first recall that sometimes long-run decisions by oligopolistic firms are taken into ac-

---

[12] Decisions on non-price competition variables are *long-run*, as compared with *short-run* decisions on prices and quantities, because they normally show a lower degree of reversibility. Thus, when a price decision is to be taken, firms find themselves in a given environment which reflects previous choices on non-price competition variables and which cannot costlessly be altered in the short-run.

[13] See Phlips (1995), Scherer and Ross (1990), Symeonidis (1997a, 1997b).

count by antitrust analysis as "facilitating practices", that is as patterns of conduct that can help collusive co-ordination. Hence, under certain conditions, a facilitating practice may be attacked for purposes of antitrust analysis as circumstantial evidence of collusive behaviour. From a quite different perspective I am instead suggesting that competition policy might look at the long-run decisions, not with the purpose of detecting and condemming them as an instance of collusive co-ordination, but because of their informative value in telling whether firms expect that a collusive equilibrium behaviour will or will not follow in the market.

Theoretical results on semicollusion can be organised along two different strands of analysis. First, one may refer to a class of models in which the choices made by oligopolists in the first stage are interpreted as directed at improving (or, indeed, maximising) the firms' relative bargaining position in the second stage. In these models the specific collusive solution is supported by the long-run decisions, in the sense that those variables are strategically selected by firms having in mind the particular collusive solution that will ensue. A second class of models ignores the way in which long-run choices can support the short-run collusive solution, which is entirely taken as exogenous. Instead, these models look at the effect on the long-run variables of dissipating the larger amount of rents that accrue to the firms thanks to the collusive behaviour in the short-run. Since both classes of models support similar predictions, I will longer dwell on the former, which provides us with a larger set of results that include long-run decisions on location (or, more generally speaking, horizontal product differentiation), capacity and R&D expenditures.

Starting with location choices, we have a general result according to which the degree of horizontal product differentiation in the first stage is negatively correlated with expectations of price-collusion. Such a result holds independently of whether collusion is explicitly or implicitly reached. In a model of explicit collusion, where the Nash bargaining solution is selected as the outcome of the second stage, price collusion leads to minimum differentiation if monetary transfers are assumed away[14]. Even when monetary transfers are possible, although there is some differentiation, products are less differentiated than they would have been, had collusion prevailed also in the first-stage.

In the standard, unit-length, Hotelling model with two firms and quadratic costs, results can be summarised as follows: (i) maximum product differentiation [i.e., location at $(0,1)$] ensues when competitive conditions prevail in both the first and the second stage: the well-known intuition being that price competition is at most relaxed when firms are located at extreme

---

[14] See Jehiel (1992).

points; ii) location at (1/4, 3/4) ensues when collusion prevails in both the first and the second stage; here the intuition being that, by so doing, firms maximise the consumer's total surplus, from which the collusive joint profits are extracted; iii) location somewhere in between (1/4 and 1/2) and (1/2 and 3/4) ensues under semicollusion because two forces are at work, provided that monetary transfers are possible: (1) choosing a product closer to the competitor's improves the firm's bargaining power; (2) choosing a product closer to the one that maximises the joint profit improves the global efficiency of the solution. Hence, the fundamental intuition that governs location choice under explicit semicollusion is that, by selecting closer products, each firm improves its bargaining strength in the following collusive agreement. The same intuition holds under tacit collusion when the enforcement conditions for the collusive behaviour must be endogenously determined: for, sameness of products means that the firms' ability to punish one another for defection is in fact maximised once the equilibrium locations are selected[15].

Turning to choice of capacity, a general result of overinvestment is obtained when decisions about capacity are taken in a semicollusive setting. Again, the different scenarios can be summarised as follows. Firstly, absent collusion in both the first and the second stage, capacity would be restricted to the level that guarantees the attainment of the Cournot outputs as a means to relax harsh price competition[16]. Secondly, since holding excess capacity is costly, capacity would be restricted to the level corresponding to the collusive output, were collusion to prevail in both stages. Finally, excess capacity will result in a semicollusive setting[17]: here again, the intuition is that colluding firms have an incentive to carry excess capacity since, by so doing, they make their threats to act competitively more damaging, thus obtaining a more favourable agreement. Although this last result was formally proved in a model in which the second stage is a Nash bargaining game, i.e. under condition of an explicit agreement, it intuitively extends to tacit collusion, since recourse to a punishing strategy can only be made credible, provided firms have enough capacity to carry it on.

Similar results of excess investment obtain when the third long-run choice variable, i.e. expenditures on R&D, is taken into account. In addition to proving that the semicollusion involves excess capacity and higher investment in R&D in comparison with the competitive equilibrium, one can also show that semicollusive profits will, under given circumstances, be smaller than the non-co-operative equilibrium profits[18].

---

[15] See Friedman and Thisse (1993).
[16] This is the well-known result due to Kreps and Scheinkman (1983).
[17] See Osborne and Pitchik (1987).
[18] See Fershtman and Gandal (1994).

Let me now briefly turn to the second class of models that look at semicollusion from a different perspective. In these models the focus is on long-run expenditures on R&D as well as on advertising as endogenously determined sunk costs. The result of excess investment on these expenditure variables under collusion in the second stage of the game still obtains even if the firms' short-run conduct is left as exogenously given. The basic intuition under this setting is that, a lower intensity of price-competition brings forth a higher level of profits gross of sunk costs on advertising and R&D, thus increasing the firms' incentive to spend on these long-run variables, up to the point where the larger rents are entirely dissipated[19].

As a final remark let me stress a striking common feature of all the results surveyed so far. When long-run decisions are taken strategically, with the purpose of strengthening collusion in the short-run, outcomes always lay outside the polar full collusion/non collusion range[20]. This striking common feature may provide a clear and interesting starting point for the necessary empirical assessment of semicollusive equilibria in an olipolistic context. Notice also the poor welfare results in the semicollusive setting: either product standardisation, or excessive investment in capacity or in R&D, are to be expected. Thus policy makers should particularly care about the welfare consequences of semicollusive behaviour.

4. Conclusions

Game-theoretical models of rational behaviour in social situations suggest that the issues of *co-ordination* and *co-operation* should be kept analytically distinct in the assessment of oligopolistic collusion for antitrust purposes.

Whereas co-ordination should not always per se be viewed as an evidence of agreement, or concerted practice, on *monopolistic pricing*, substantial, i.e. economic, analysis of collusive co-operation should always be in order, even if evidence of direct co-ordination on the price is lacking. To this purpose, enlarging the observed set of the firms' decision variables can provide useful insights. In particular, expectations of collusive behaviour in the short-run significantly alter the oligopolistic firms' decisions on long-run choices. Thus, the observation of the firms' long-run choices (e.g., in terms of unwarrantable failure of horizontal product differentiation or of excess investment in capacity and in R&D) should have a say in inferring a possible underlying collusive equilibrium in the market. It is worth noticing that the multistage game leaves adequate room for accommodating the legal preoccupation that the parties' common understanding be given distinct and separate evidence from the collusive conduct itself, since the firms' choices

---

[19] See Sutton (1991) and Symeonidis (1997a, 1997b).

[20] See Friedman and Thisse (1993).

in the first stage are to be explained as a result of the intent to collude. Moreover, if long-run choices are taken as an evidence of ensuing collusive conduct, there would be an incentive for firms to modify their behaviour, in order that collusion cannot be inferred. Thus, in most circumstances, namely when product differentiation and capacity choices are involved, the equilibrium in the second stage would be beneficially affected, since the incentives to collude on the price would be eroded if decisions on these long-run variables are modified.

Though the results that I have surveyed are promising, I must conclude with a note of warning: the models to which I have referred are descriptive models and their results lack, as they now stand, a definite normative conclusion. By this I mean that, while we have theorems based on "if" conditions - such as "provided that collusion ensues in the short run, then long-run decisions will be such and such", - we still are strongly in need of theorems based on "only if" conditions. Hence, in absence of further work the results surveyed are to be handled with care. Indeed, further work is strongly needed. Competition Authorities can have at their disposal a huge amount of information and data about firms' strategies in oligopoly: it really is regrettable that this information is doomed to remain pretty useless, as long as we are bound to only rely on possibly deceptive evidence of co-ordination on price strategies or on evidence of information exchange.

References

Baker J. (1993), *Introduction to Symposium on Tacit Collusion*, The Antitrust Bulletin, vol. XXXVIII, Spring, p.1 ff.

Fershtman C. and N. Gandal (1994), *Disadvantageous semicollusion*, International Journal of Industrial Organisation, vol., 12, p.141 ff.

Friedman J. and J.F. Thisse (1993), *Partial collusion fosters minimum product differentiation*, Rand Journal of Economics, vol.24, p.631 ff.

Jehiel P. (1992), *Product differentiation and price collusion*, International Journal of Industrial Organisation, vol.10, p. 633 ff.

Kreps D. and J. Scheinkman (1983), *Quantity Precommitment and Bertrand Competition Yield Cournot Outcomes*, Bell Journal of Economics, vol.14, p.326 ff.

Osborne M.J. and C. Pitchik (1987), *Cartels, Profits and Excess Capacity*, International Economic Review, vol.28, p.413 ff.

Phlips L. (1995), *Competition Policy: a Game-Theoretic Perspective*, Cambridge University Press.

Scherer F.M. and D. Ross (1990), *Industrial Market Structure and Economic Performance*, Boston, Houghton Mifflin.

Sutton J. (1991), *Sunk Costs and Market Structure*, Cambridge, MIT Press.

Symeonidis G. (1997a), *Price Competition and Market Structure: The Impact of Restrictive Practices Legislation on Concentration in the UK*, STICERD Working Paper n.EI/18, London School of Economics.

Symeonidis G. (1997b), *Cartel Policy, Non-Price Competition and Market Structure: Theory and Evidence from the U.K.*, STICERD Working Paper n.EI/19, London School of Economics

Turner D. (1962), *The Definition of Agreement Under the Sherman Act: Conscious Parallelism and Refusals to Deal*, Harvard Law Review, n.75, p. 655 ff.

Weber M. (1922), *Economy and Society*, G. Roth and P. Wittich (eds.), New York, Bedminster Press, 1968.

Whish R. and B. Sufrin (1993), *Oligopolistic Markets and the EC Competition Law*, Yearbook of European Law n.12, Clarendon Press, Oxford, p.59 ff.

Yao D. and S. De Santi (1993), *Game Theory and the Legal Analysis of Tacit Collusion*, The Antitrust Bulletin, vol. XXXVIII, Spring, p.113 ff.

CHAPTER 10

STRUCTURAL ESTIMATION OF AUCTION MODELS

HAN HONG (doubleh@princeton.edu)
Princeton University
Department of Economics
Fisher Hall, Princeton, New Jersey 08544-1021, USA

MATTHEW SHUM (mshum@chass.utoronto.ca)
University of Toronto
Department of Economics
150 St. George Street, Toronto, Ontario Canada M5S 3G7

Abstract. This paper surveys the existing literature on and considers a general methodological approach for the structural estimation of empirical auction models, in which
the estimating equations are directly derived from the equilibrium bid functions posited
in the theoretical literature. We describe the building blocks of an empirical structural
auction model, and illustrate using the first-, second-, and ascending-auction models as
examples. Alternative estimation approaches for these models are discussed.

1. Introduction

Since the seminal work of William Vickrey, auction theory has developed
into one of the most sophisticated and systematically investigated liter-
atures in economics. Furthermore, auctions are being increasingly used
as real-world allocation mechanisms (the most celebrated example being
the spectrum auctions run the U.S. Federal Communications Commission), thereby raising many interesting empirical issues. This combination
of highly-developed theory and real-world applications has spawned an
impressive body of empirical work.

This paper examines one branch of this empirical work — that dealing with structural estimation of auction models. This work (see Paarsch
(1992), Paarsch (1991) for two of the earliest examples) derives the estimating equations directly from the equilibrium bid functions posited in the
theoretical auction literature, and attempts to recover the parameters of

the underlying distribution of bidders' valuations. In contrast, the *reduced-form* empirical auction literature (see Hendricks and Porter (1988) for an example) tests the comparative statics predictions of the theoretical auction models, without directly recovering the parameters of the distribution of bidders' valuations.

These two approaches are used to address different types of questions. The reduced-form approach, which aims more to characterize bidder behavior in an auction rather than to use the equilibrium bid functions as a mapping from observed bids to (unobserved) bidder valuations in order to estimate the parameters of the distribution of the latter. For example, the analysis of Hendricks and Porter (1988) tries to uncover patterns between the observed bids, the number of participating bidders, and (proxies of) differences in informedness among bidders in offshore oil and gas auctions, patterns which are predicted by auction theory.

On the other hand, the structural approach explicitly recovers the parameters of the distribution of bidders' valuations. These parameters allow the researcher to simulate auction results under alternative auction formats, which is crucial for comparing the efficiency and seller revenue optimality of alternative auction forms, as well as evaluating the effects of policy changes. For example, Paarsch (1991) uses his estimates to calculate the optimal reserve prices for British Columbian timber auctions, an important source of government revenue. The price in taking the structural approach is the extra assumptions that the researcher must make, relative to the reduced-form approach. Since, by definition, the estimating equation in a structural auction model is derived from the equilibrium bid functions of a theoretical auction model, the researcher must make the same assumptions that are made in the corresponding theoretical model. Furthermore, structural estimation is impossible for auction formats for which the forms of the equilibrium bid functions are not known.

This paper offers a general view of structural estimation using auction data, and emphasizes the common components behind most structural econometric auction model. In the next section, we discuss the general framework of a structural auction model. Section 2 illustrates this framework with several examples from the literature. Section 3 discusses estimation methodologies and the econometric problems encountered in the estimation procedure. Section 4 concludes.

2. A general structural empirical auction model

We restrict our attention to single object auctions. An auction has N bidders (indexed $i = 1, \ldots, N$), each of whom have a valuation V_i for the object, and receive a private signal X_i about V_i. Bidder i only observes

X_i prior to the beginning of the auction. He doesn't observe any of the valuations, V_j, for $j = 1, \dots, N$, or any of the other bidder's signal, X_j, for $j \neq i$.

The bidders' valuation s and private signals are jointly distributed according to the distribution function $F(V_1, \dots, V_N, X_1, \dots, X_N)$. The researcher estimates a structural model in order to identify the F distribution, or parameters thereof.

An auction model is distinguished by (i) the *form* of the auction being studied and (ii) the assumptions underlying the joint distribution of the bidders' valuations and their signals, which we call the *paradigm* of the auction.

The form of an auction includes the bidding rules (e.g. sealed-bid vs. open-cry, whether there is a reserve price) as well as the allocation rule which dictates what price the winning bidder must pay for the object. The most commonly used auction forms are first-price auctions, second-price auctions, Dutch auctions and English(ascending) auctions. Particulars about these auction forms are given in Milgrom and Weber (1982).

For the set of assumptions made regarding $F(V_1, \dots, V_N, X_1, \dots, X_N)$, we refer to them as the "paradigm" of a particular auction model. In the pure *private value* paradigm, $V_i = X_i \; \forall \; i$ (i.e. each bidder knows his true valuation for the object) while in the pure *common value* paradigm $V_i = V, \; \forall \; i$ (i.e. the value of the object is the same to all bidders, but none of the bidders knows the true value of the object; here the individual X_i's are noisy signals of the true but unknown V). Typically, however, we would expect that there are both private and common value components in the valuation that a bidder places on the object on sale. Generally speaking, V_i is a function of *all* the bidders' signals: and other information variables that may not be observed by any of the bidders.

When the joint distribution F is symmetric with respect to $1, \dots, N$, the model is *symmetric*. Otherwise it is *asymmetric*. When the joint distribution F can be factored into marginal distributions of $V_i, X_i, i = 1, \dots, N$ (i.e., $F(V_1, \dots, V_N, X_1, \dots, X_N) = F_1(V_1, X_1) \, F_2(V_2, X_2) \, \dots \, F_N(V_N, X_N)$), the bidders' valuations are labeled *independent*. Different combinations of private/common value assumptions, symmetric/asymmetric distribution assumptions and independence/dependence assumptions create a host of possible paradigms for each auction form. Milgrom and Weber (1982) provided the seminal analysis for the symmetric versions of most of the usual auction forms; much of the recent theoretical work has focused on asymmetric cases (Maskin and Riley (1996), Bulow, Huang and Klemperer (1996)).

The data are the observed bids $p_1, \dots, p_N$. In many applications the researcher only observes the winning bid. In other applications, the researcher observes all the bids, but may choose only to estimate using winning bid data.

Given the assumptions underlying each model paradigm, a structural empirical auction model has two components:

1. Equilibrium bid functions Theoretical equilibrium characterization results contribute the form that the equilibrium bidding strategies take. Bidder i's equilibrium bid is a function of all the information variable available to him, denoted by Ω_i. In a sealed bid auction, the only information variable he has is his own private signals, X_i. On the other hand, in an irreversible dropout ascending auction[1], the information variables available to him during a stage of the auction include all the signals of the already dropped-out bidders, which bidder i infers from the dropped-out prices.

Generally, bidder i's equilibrium bid can be considered a function of all the private signals: $b_i(X_1, \ldots, X_N)$, with the understanding that the form of the bid function $b_i(\cdots)$ depends both on the auction form, and on F, the joint distribution of valuation and signals. For example, in seal bid auctions, $b_i(\cdots)$ depends only on X_i. The collection of equilibrium bid functions $b_1(\cdots; F), \ldots, b_N(\cdots; F)$ provide a mapping between the private signals $X_1, \ldots, X_N$ and the observed bids $p_1, \ldots, p_N$. Monotonicity assumptions about the equilibrium bid functions usually ensure that this mapping is one-to-one.

2. Researcher's distribution assumptions regarding **F** Given this mapping, the distributional of the signals X's induce a joint distribution function for the observed bids $G(p_1, \ldots, p_N)$, usually in the form of a joint density function $g(p_1, \ldots, p_N)$, which forms the basis for an estimating equation. An interesting feature of the structural auction model is that $G(p_1, \ldots, p_N)$ depends on F in two ways. Consider $g(p_1, \ldots, p_N) = f_X(X_1, \ldots, X_N) \times J(X_1, \ldots, X_N | p_1, \ldots, p_N)$, where $J(\cdots)$ denotes the Jacobian transformation from the private signals to the observed prices. $f_X(\cdots)$ is simply given by the marginal joint density of $X_1, \ldots, X_N$ from F. On the other hand, $J(\cdots)$ usually also depends on F, in the sense that for the same auction form, j will vary depending on assumptions made about F.

The joint distribution of observed bid data, $G(p_1, \ldots, p_N; F)$ provides a way of identifying the underlying latent distribution F (which is the ultimate goal of structural estimation of auction models) from the observed bids. In the following we will focus on parametric models, in which the form of F is assumed to be known up to a finite dimensional vector of parameters. All the information needed for any estimation procedure is summerized

---

[1] the form considered in Milgrom and Weber (1982) and in most subsequent work on ascending auctions

in $g\left(\cdot\right)$, the joint density of the bids. In maximum likelihood estimation procedure, we use $g(\cdots)$ directly as the estimating equation. This is the approach taken by in the Paarsch papers cited earlier. Alternatively, the moments of g can also be used to do minimum distance estimation (non-linear least squares), as in Laffont, Ossard and Vuong (1995) and Hong and Shum (1997). In what follows, we present a survey of previous work on structural estimation of auction models, fitting these examples into the general framework described above.

3. Deriving distribution of observed bid data: some examples

3.1. FIRST-PRICE AUCTION MODELS

The first-price auction proceeds as follows: Observing $X_i = x$, bidder i chooses a bid b_i to maximize his expected payoff, given the other bidders' equilibrium behavior:

$$b_i = \text{argmax}_b E\left[(V_i - b)\, 1\left(X_j \le b_j^{-1}\left(b\right), j \ne i\right) | X_i = x\right]$$

where as $b_i\left(\cdot\right), i = 1,\dots,n)$ denotes the equilibrium bidding strategy (or *bid function*) for bidder i.

The first order condition of this maximization problem is:

$$\sum_{j \ne i} \left\{ \frac{\partial F_i\left(b_j^{-1}\left(b\right), j \ne i | X_i = x; \theta\right)}{\partial X_j} \frac{1}{b_j'\left(b_j^{-1}\left(b\right)\right)} \right.$$

$$E\left(V_i | X_i = x, X_j = b_j^{-1}\left(b\right), X_k \le b_k^{-1}\left(b\right), k \ne i, j\right) \bigg\}$$

$$-F_i\left(b_j^{-1}\left(b\right), j \ne i | X_i = x; \theta\right) - b \sum_{j \ne i} \frac{\partial F_i\left(b_j^{-1}\left(b\right), j \ne i | X_i = x; \theta\right)}{\partial X_j} \frac{1}{b_j'\left(b_j^{-1}\left(b\right)\right)} = 0$$

$$(1)$$

where $F_i\left(X_{-i} | X_i; \theta\right)$ denotes the conditional distribution of X_{-i}, the $N-1$ subvector of the signals excluding X_i, given $X_i = x$.

The system of N first-order-conditions implicitly defines the set of N equilibrium bid functions $b_1,\dots,b_n$. These equations simplify under certain assumptions, which we now proceed to make in steps.

First, for a priate value model ($V_i = X_i, \forall i$), $E(V_i \mid \cdots) = X_i$, so that the system of differential equations simplify to:

$$(X - b_i\left(X\right)) \sum_{j \ne i} \frac{h_j\left(b_j^{-1}\left(b_i\left(X\right)\right)\right)}{H_j\left(b_j^{-1}\left(b_i\left(X\right)\right)\right)} \frac{1}{b_j'\left(b_j^{-1}\left(X\right)\right)} = 1, \quad \text{for } i = 1,\dots,N \quad (2)$$

with the boundary conditions $b_i(\underline{x}) = \underline{x}$, where $\underline{x}$ is the lower bound of the support of a signal X.

Bajari (1996) relies on computational procedures to solve the system of differential equations in (2). In a nonparametric framework, Vuong, Perrigne and Guerre (1996) showed that it is possible to nonparametrically identify F from all bid data for private value first-price auction models, regardless of the asymmetry and dependence of F, as long as F is such that it gives rise to a strictly increasing strategy equilibrum.

Next, if we assume the signals to be independently distributed, the joint density of the observed bids (applying to both symmetric and asymmetric models) becomes $g(p_1,\dots,p_N) = \prod_{i=1}^{N} \frac{h_i(b_i^{-1}(p_i))}{b_i'(b_i^{-1}(p))}$

Finally, Laffont, Ossard and Vuong (1995) made the additional assumption of symmetry. This implies that $F_X(x_1,\dots,x_N) = \prod_{i=1}^{N} H(x_i)$. In these models, the equilibrium bid functions are defined by a first-order differential equation

$$b'(x)F(x)^N + Nb(x)f(x)F(x)^{N-1} = Nxf(x)F(x)^{N-1}$$

together with the boundarty condition that $b(p_0) = p_0$. Given the symmetry assumption, the bid function is expressible analytically as:

$$b(X_i) = X_i - \frac{1}{(H(X_i))^{N-1}} \int_{p_0}^{X_i} (H(x))^{N-1} dx \qquad (3)$$

where p_0 is the reservation price in the auction. Without simplifying parametric assumptions (see Paarsch (1992) for examples of several), no general closed-form solution exists for g, given the nonlinearity in the transformation from x_i to p_i expressed in equation (3).

3.2. SYMMETRIC INDEPENDENT PRIVATE VALUE SECOND-PRICE ("VICKREY") AUCTION

This is the case considered by Paarsch (1991). Here $F(X_1,\dots,X_N) = \prod_{i=1}^{N} H(x_i)$, and $b_i = X_i$, $\forall i$ is unique increasing dominant-strategy equilibrium. Given these strategies, the joint density of observed bids is the same as (assumed) joint density of the bidder signals:

$$g(p_1,\dots,p_N) = h(p_1)h(p_2)\dots h(p_N) \qquad (4)$$

Paarsch assumes that H is the Weibull distribution, a flexible two-parameter distribution which can have a monotonically increasing or decreasing hazard rate depending on the values of the parameters.

Given the simple form of the equilibrium bidding strategies in the second-price private value auction, if only the winning bid is observed, its density would be that of the second highest draw out of N draws from the H distribution:

$$f_{X(2:N)}(X) = N(N-1)[H(X)][1 - H(X)]^{N-2}h(X) \qquad (5)$$

3.3. ASYMMETRIC OPEN AUCTION MODELS

3.3.1. *Private value models*

Ever since Vickrey (1962), it has been known that in private value second-price auctions and ascending auctions (which we will jointly refer to as **open auctions**), bidding up to the private value $(b_i(X_i) = X_i)$ is the unique weakly undominated strategy equilibrium, *regardless of symmetry or independence assumptions.*[2] Therefore, the Vickrey auction provides an example of an auction model where the Jacobian of the transformation from the unobserved X's to the observed p's does not explicitly depend on F, the joint distribution function for the X's. In equilibrium, $p_i = X_i$, so the Jacobian is simply the identity matrix. This property disappears once common value components are added to the model. This is the focus of the next section.

3.3.2. *Models with common values*

In common value auction models, it is assumed that there is an (unknown to all bidders) component in bidders' valuations of the object which is the same (i.e., "common") across all bidders. For this reason, each private signal X_i is useful to each bidder $j \neq i$ in estimating his valuation V_j, so that V_j is typically a function of *all* the private signals $X_1, \ldots, X_N$.

Note the subtle difference between a common value model and a model with correlated private values (such as that considered in Vuong, Perrigne and Li (1997)). Both models assume correlation among the private signals, i.e., $F_X(X_1, \ldots, X_N)$ cannot be factored into $\prod_i H_i(X_i)$. In the correlated private value model, it remains the case that $V_i = X_i$, for $i = 1, \ldots, N$. In the common value model, however, $V_i = v_i(X_1, \ldots, X_N)$, for $i = 1, \ldots, N$, which differs from the private value case.

As we will show, equilibria in open auctions with common value components under both asymmetry and dependence have a very intuitive derivation, even allowing for asymmetry and dependence in bidders' private signals. One striking finding is that the *inverse* bid functions, i.e., what a bidder's signal would be if he chooses to bid the given price level p, can be derived as solutions to a system of nonlinear equations defined as bidders'

---

[2] In fact, Vickrey generalizes this "highest rejected bid" principle to the simultaneous auctions of identical objects.

expected valuations conditional on all the information available during that stage of the auction. From a computational point of view, this provides a way to derive the mapping from the unobserved X's to the observed p's numerically in the usual case when it is analytically intractable.

Here let us introduce the shorthand notation that bidder i's bid function be written simply as a function of his signal x_i and the public information set Ω: $b_i(x_i) \equiv b_i(x_i; \Omega)$. Then in these open auctions:

$$b_1(x_1) = E[V_1 \mid \mathcal{A}_1(x_1, b_2(X_2), \dots, b_N(X_N)), \Omega]$$

$$\dots \quad (6)$$

$$b_N(x_N) = E[V_N \mid \mathcal{A}_N(b_1(X_1), \dots, B_{N-1}(X_{N-1}), x_N), \Omega]$$

where $\mathcal{A}_i(\cdots)$ denotes the conditioning event for bidder i. Typically, $\mathcal{A}_i$ involves bidder i's private signal X_i as well as his equilibrium beliefs about the other bidders' signal $X_1, \dots, X_{i-1}, X_{i+1}, \dots, X_N$. Furthermore, as will be seen below, $\mathcal{A}_i$ relates all the other bidders' signals to bidder i's signal through the equilibrium bid functions $b_i(X_i)$, $i = 1, \dots, N$. In what follows we will explicitly write out $\mathcal{A}_i$ for the second price and ascending auctions.

For a generic bid $p = b_i(x_i)$, $i = 1, \dots, N$, the special form of A_i allows us to rewrite the above system of conditional expectations as a system of N nonlinear equations, with N unknowns $\phi_1(p), \dots, \phi_N(p)$, where $\phi_i(p) \equiv b_i^{-1}(p)$, the inverse bid function for bidder i evaluated at the generic bid p; i.e. the system (6) can be rewritten as

$$p = E[V_1 \mid \mathcal{A}_1(\phi_1(p), \dots, \phi_N(p)); \Omega]$$

$$\vdots \quad (7)$$

$$p = E[V_N \mid \mathcal{A}_N(\phi_1(p), \dots, \phi_N(p)); \Omega].$$

By taking different values for p, we can solve the system (7) for the N inverse bid functions in pointwise fashion. Given any distribution F then, the existence, uniqueness, and monotonicity properties of the equilibrium bid functions can be directly verified from the existence, uniqueness, and monotonicity of solutions to the system of equations (7).

Note that assumptions about F, the joint distribution of bidders' private signals, determine in practice only the form that the conditional expectations will take. The form that the equilibrium bid functions take depends on the nature of the solution to the nonlinear system of equations posed by the conditional expectation equations. It is in this sense that J, the Jacobian of the mapping from the unobserved X's to the observed p's depends explicitly on the assumptions made regarding F. This is not the case for limited dependent variable (LDV) models where the rules for mapping the unobserved utility indices to the observed LDV's are threshold-crossing conditions which are invariant to the distribution assumed for the unobserved utility indices.[3]

Even given parametric assumptions about F, it is rare to find cases where the conditional expectations will have a closed form, much less cases where both the conditional expectations and the equilibrium bid functions are expresible in closed form. An example of this is the log-normal irreversible-dropout English auction model, examined by Wilson (1995) and recently implemented empirically by Hong and Shum (1997).

In general, solving for the equilibrium bid functions will require numerical procedures both at the conditional expectation evaluation stage (requring numerical integration) and the stage of solving for the implicitly-defined equilibrium bid functions (function approximation routines).

3.3.3. Asymmetric second price auctions

For general asymmetric second-price auctions, the equilibrium bidding strategies consist of one bid function per bidder, i.e. a set of functions $b_i(X_i)$, for $i = 1, \ldots, N$. In equilibrium, bidder i believes that his bid is equal to the highest competing bid, i.e., $\max_{j \neq i} b_j(X_j) = b_i(X_i)$. Therefore, the equilibrium bid functions satisfy the following system of conditional expectations:

$$b_1(X_1) = E[V_1 \mid X_1, \max_{j \neq 1} b_j(X_j) = b_1(X_1)]$$

$$b_2(X_2) = E[V_2 \mid X_2, \max_{j \neq 2} b_j(X_j) = b_2(X_2)]$$

$$\cdots$$ (8)

$$b_N(X_N) = E[V_N \mid X_N, \max_{j \neq N} b_j(X_j) = b_N(X_N)].$$

Next we will rewrite these equations using the inverse bid functions, analogous to the system (7) above.

Define $\beta_{-i} \equiv \max_{j \neq i} b_j(X_j)$ and the function $\tilde{V}_i(x, p) \equiv E(V_i \mid X_i = x, \beta_{-i} = p)$. We will first discuss the event $\{\beta_{-i} = p\}$. $\{\beta_{-i} = p\}$ means that the highest bid among bidders $j, j \neq i$ is p, which in turn implies that

§1 All bids by bidders $j, j \neq i$ are smaller than or equal to p. $b_j(X_j) \leq p$, $\forall \, j \neq i$.
§2 At least one of $b_j(X_j) = p, j \neq i$.

Therefore we can write the event $\{\beta_{-i} = p\}$ as

$$\left[\bigcap_{j \neq i} (b_j(X_j) \leq p) \right] \cap \left[\bigcup_{j \neq i} b_j(X_j) = p \right] = \left[\bigcap_{j \neq i} (X_j) \leq \phi_j(p) \right] \cap \left[\bigcup_{j \neq i} (X_j) = \phi_j(p) \right]$$ (9)

---

[3] Strictly speaking, the Jacobian is not defined from most LDV models. However, the rule for transforming latent utility into observed actions does not depend on model parameters.

where $\phi_j(p) = b_j^{-1}(p)$ are the inverse bid functions of bidders $j, j \neq i$.[4] In view of (9), we can rewrite (8) as:

$$p = E\left(V_1 \middle| X_1 = \phi_1(p), \left[\bigcap_{j \neq 1}(X_j) \leq \phi_j(p)\right] \cap \left[\bigcup_{j \neq 1}(X_j) = \phi_j(p)\right]\right)$$

$$\vdots \quad (10)$$

$$p = E\left(V_N \middle| X_N = \phi_N(p), \left[\bigcap_{j \neq N}(X_j) \leq \phi_j(p)\right] \cap \left[\bigcup_{j \neq N}(X_j) = \phi_j(p)\right]\right)$$

which, analogously to the system (7), is a system of N equations in the N unknowns $\phi_1(p), \ldots, \phi_N(p)$ which can be solved for different p's for the inverse bid functions $\phi_1, \ldots, \phi_N$.[5]

3.3.4. Asymmetric ascending auctions
The ascending auction proceeds in rounds. It enters a new round whenever another bidder drops out. N bidders are present in the auction; there will be $N - 1$ "rounds" in the auction, indexed $k = 0, \ldots, N - 2$. In round 0, all N bidders are active; in round k, only $N - k$ bidders are active. Each round ends when a bidder drops out; bidders are indexed by $i = 1, \ldots, N$.

---

[4] Because we assume that the type space is continuous, the event (9) has zero probability. However, the conditional expectation we are computing is well defined as long as we assume $f(v, x) > 0$ on its support.

[5] In calculating the conditional expectations in (10), we will express it as:

$$\check{V}_i(x_1, \ldots, x_N) = E\left(V_i \middle| X_i = x_i, \bigcup_{j \neq i}[X_j = x_j, X_k < x_k, k \neq j, i]\right)$$

$$= \frac{\int V \sum_{j \neq i} \int^{x_{-ij}^1} \cdots \int^{x_{-ij}^{n-2}} f_{V_i, X_1, \ldots, X_N}\left(V, x_i, x_j, X_{-ij}^1, \ldots, X_{-ij}^{N-2}\right) dX_{-ij}^1 \cdots dX_{-ij}^{N-2} dV}{\sum_{j \neq i} \int^{x_{-ij}^1} \cdots \int^{x_{-ij}^{N-2}} f_{X_1, \ldots, X_N}\left(x_i, x_j, X_{-ij}^1, \ldots, X_{-ij}^{N-2}\right) dX_{-ij}^1 \cdots dX_{-ij}^{N-2}}$$

$$(11)$$

where X_{-ij} denotes the $N-2$ vector of private signals for bidders other than i and j, and X_{-ij}^k denotes the generic kth element of the vector X_{-ij}. Each term in the summation presents one event in the union; they are disjointed from each other. The $N - 2$ integral in the denominator is the joint density of the conditioning event. It integrates over the signal of each of the bidders, other than i and j, from the lower bound $\underline{X}$ up to each of the $x_k = \phi_k(b)$. Unless the integrals can be analytically expressed, in general it can be computationally intensive to calculate this conditional expectation because it involves multi-dimensional integrals.

In particular, even for jointly normally distributed (v, x), the difficulty of evaluating the multivariate normal distribution function is similar to that encountered in the estimation of multivariate probit models. One recently proposed solution of this problem is to use simulation estimators, which evaluates the integral by the empirical average from many independent random draws (see, for example, Hajivassiliou and McFadden (1998) and Hajivassiliou and Ruud (1994)).

Without loss of generality, the ordering $1, \dots, N$ indicates the order of dropout. In other words, bidder N drops out in round 0, and bidder 1 wins the auction; generally, bidder $N - k$ drops out at the end of round k. The dropout prices are indexed by rounds, i.e. $P_0, \dots, P_{N-2}$.[6] To sum up, bidder i drops out at the end of round $N - i$, at the price P_{N-i}.

Equilibrium bidding strategies in the ascending auction game specify, for each bidder i, bid functions $b_i^k(X_i)$ for each round k, $k = 0, \dots, N - 2$, i.e. $b_i^0(X_i), \dots, b_i^{N-2}(X_i)$. Given a realization of the private signal X_i, the bid function $b_i^k(X_i)$ tells bidder i which price he should drop out at during round k. The collections of bid functions $b_i^0(X_i), \dots, b_i^{N-2}(X_i)$ for bidders $i = 1, \dots, N$ are common knowledge. The equilibrium conjectures A_i and the bidders' expectations (6) evolve during different rounds of the auction.

Again, consider bidder i, who is active during round k. As of round k, bidders $N - k + 1, \dots, N$ have already dropped out, at prices $P_{k-1}, \dots, P_0$, respectively. Since the equilibrium bid functions are common knowledge, bidder i can use this information on the identity of the dropout bidders and their dropout prices to infer the private signals $X_{N-k+1}, \dots, X_N$ observed by these bidders by inverting these bid functions: $X_j = (b_j^{N-j})^{-1}(P_{N-j})$, for $j = N - k + 1, \dots, N$.

The price p at which bidder i should quit the auction during round k, defined as his *bid function* for round k, is the price $b_i^k(X_i) \equiv p$ at which he will have a zero expected profit in round k if all other active bidders simultaneously quit at the same price. In equilibrium, the conditioning event $\mathcal{A}_i^k$, which changes for a given bidder across rounds (therefore the superscript k) consists of (1) bidder i's private signal X_i; (2) the private signals of the bidders who have dropped out before round k, $X_{N-k+1}, \dots, X_N$, where $X_j = (b_j^{N-j})^{-1}(p_{N-j})$, for $j = N - k + 1, \dots, N$; and (3) bidder i's beliefs that all the other remaining bidders have the same targeted dropout price as he:

$$b_1^k(X_1)$$
$$= E\left[V_1 \mid X_1, b_j^k(X_j) = b_1^k(X_1), j = 2, \dots, N - k, \Omega_k\right]$$
$$b_2^k(X_2)$$
$$= E\left[V_2 \mid X_2, b_j^k(X_j) = b_2^k(X_2), j = 1, 3, \dots, N - k, \Omega_k\right] \qquad (12)$$

$$\dots$$

$$b_{N-k}^k(X_{N-k})$$
$$= E\left[V_{N-k} \mid X_{N-k}, b_j^k(X_j) = b_{N-k}^k(X_{N-k}), j = 1, \dots, N - k - 1, \Omega_k\right]$$

---

[6] Note that P_{N-1}, the winner's bid, will generally not be observed in ascending auction datasets.

where $\Omega_k \equiv \left\{X_j = (b_j^{N-j})^{-1}(P_{N-j}), \text{ for } j = N - k + 1, \ldots, N\right\}$. The full
set of equilibrium bid functions is analogously described by sets of $N - k$
equations for each round $k = 0, \ldots, N - 1$.

In equilibrium, this entire system of equations must hold for any bid p,
and the set of signals $\phi_i^k(p) \equiv (b_i^k)^{-1}(p)$, for $i = 1, \ldots, N - k$. If we treat p
as a parameter and the inverse bid functions $\phi_i^k)(p)$, $i = 1, \ldots, N - k$ as
the unknown variables, we can rewrite (12) as a system of $N - k$ equations
in $N - k$ unknowns, analogous to (7):

$$p = E\left[V_1 \mid X_1 = \phi_1^k(p), \ldots, X_{N-k} = \phi_{N-k}^k(p); \right.$$
$$\left. X_{N-k+1} = \phi_{N-k+1}^{k-1}(p_{k-1}), \ldots, X_N = \phi_N^0(p_0)\right]$$
$$p = E\left[V_2 \mid X_1 = \phi_1^k(p), \ldots, X_{N-k} = \phi_{N-k}^k(p); \right.$$
$$\left. X_{N-k+1} = \phi_{N-k+1}^{k-1}(p_{k-1}), \ldots, X_N = \phi_N^0(p_0)\right] \qquad (13)$$

$$\cdots$$

$$p = E\left[V_{N-k} \mid X_1 = \phi_1^k(p), \ldots, X_{N-k} = \phi_{N-k}^k(p); \right.$$
$$\left. X_{N-k+1} = \phi_{N-k+1}^{k-1}(p_{k-1}), \ldots, X_N = \phi_N^0(p_0)\right]$$

where $\phi_i^k(p) \equiv (b_i^k)^{-1}(p)$.

Looping over rounds $k = 0, \ldots, N - 2$ and for different values of p,
we can solve in pointwise fashion for the set of inverse equilibrium bid
functions $(\phi_1^{N-1}(p), \ldots, \phi_i^{N-i}(p), \ldots, \phi_N^0(p))$ which map the observed bids
$p_0, \ldots, p_{N-2}$ to the private signals by the relation $X_i = \phi_i^{N-i}(p_{N-i})$.

From a computational point of view, the structure of the round k equilibrium bid functions (13) is particularly attractive since the conditioning
events are *points* rather than *sets*, as is the case for asymmetric second
price auctions (cf. equations (10)). In the case of the latter, evaluation
of conditional expectations would involve multi-dimensional integration,
which can be cumbersome as the number of dimensions becomes large.

3.3.5. *Consistency conditions in asymmetric ascending auctions*

In deriving the joint distribution of bids, $G(p_0, \ldots, p_{N-2})$, in an asymmetric
ascending auction models, the researcher conditions on the dropout order
observed in the data. In other words, the order of the bids and the identity
of their bidders is taken as given in specifying the conditioning events in
each round.

However, it is possible that, for some parameter values θ, the bidder
signals inferred from the calculated equilibrium bid functions (i.e., the
inverse of the set of functions solved in pointwise fashion from systems of
equations like those in (13)) imply a dropout order which differs from the
observed dropout order. These signals will be inconsistent with the specified

form of the equilibrium bid functions, whichtake as given the observed bid ordering.

An example For clarification, we consider a 4-bidder example. If the bid order among 4 bidders is 2,3,1,4, then in deriving the density of bidder 3's bid we condition on this bid order in the sense of assuming that bidder 3 knows bidder 2's signal,[7] and in deriving the density of bidder 1's bid we assume that he has observed the signals of bidders 3 and 2.[8] In specifying bidder 1's equilibrium bidding strategy in round 3, for example, we assume that upon observing bidder 3's exit in the previous round, bidder 1 inverts bidder 3's equilibrium bid function for round 2 at the observed dropout price to obtain X_3, i.e., $X_3 = (b_3^2)^{-1}(p_3; \theta)$. We include θ as an argument here to make explicit the dependence of the bid function on parameters that determine the joint distribution F.

To be more specific, given knowledge of the inverse bid functions $\phi_i^k(p; \theta)$, we can recover bidders' private signals via the relations

$$
\begin{aligned}
X_1 &= \phi_1^3\,(p_1; \theta) \\
X_2 &= \phi_2^1\,(p_2; \theta) \\
X_3 &= \phi_3^2\,(p_3; \theta) \\
X_4 &= \phi_4^4\,(p_4; \theta)\,.
\end{aligned}
\tag{14}
$$

These signals could imply a different dropout order (say, 2,4,1,3) if it were the case that, given θ:

$$
b_2^1(X_2 = \phi_2^1(p_2; \theta); \theta) = \min_{i=1,2,3,4} b_i^1(X_i = \phi_i^1(p_i; \theta); \theta)
$$

i.e., bidder 2 drops out

$$
b_4^2(X_4 = \phi_4^4(p_4; \theta); \theta) = \min_{i=1,3,4} b_i^2(X_i = \phi_i^2(p_i; \theta); \theta)
\tag{15}
$$

i.e., bidder 4 drops out

$$
b_1^3(X_1 = \phi_1^3(p_1; \theta); \theta) = \min_{i=1,3} b_i^3(X_i = \phi_i^3(p_i; \theta); \theta)
$$

i.e., bidder 3 wins.

where $b_i^k(x) \equiv (\phi_i^k)^{-1}(x)$.

---

[7] i.e., $p_3 = b_3^2(X_3) = E[V_1 \mid X_3, X_1 = \phi_1^2(p_3), X_4 = \phi_4^2(p_3); \omega_2 = \{X_2\}]$, which assumes that by round 2, X_2 is already in the public information set ω_2 which bidder 3 conditions upon in forming his bid in that round.

[8] i.e., $p_1 = b_1^3(X_1) = E[V_1 \mid X_1, X_4 = \phi_4^3(p_1); \omega_3 = \{X_2, X_3\}]$, which assumes that by round 3, X_2 and X_3 are already in the public information set ω_1 which bidder 1 conditions upon in forming his bid for that round.

However, our specification of the equilibrium bidding strategies condition on the observed dropout order $(2,3,1,4)$. In other words, bidder 1's observed dropout bid p_1 is modeled as in equation (13):

$$p_1 = b_1^3(X_1) =$$
$$= E[V_i \mid X_1, X_2 = (b_2^1)^{-1}(p_2), X_3 = (b_3^2)^{-1}(p_3), X_4 = (b_4^3)^{-1}(b_1^3(X_1))] \tag{16}$$

If, in fact, given the parameter vector θ, the draw of $X_1, \ldots, X_4$ from $F_X(X_1, \ldots, X_4 \mid \theta)$ yields the bid ordering $(2,4,1,3)$, then p_1 is modeled as

$$p_1 = b_1^3(X_1) =$$
$$= E[V_i \mid X_1, X_2 = (b_2^1)^{-1}(p_2), X_3 = (b_4^3)^{-1}(b_1^3(X_1)), X_4 = (b_3^2)^{-1}(p_4)] \tag{17}$$

which is clearly inconsistent with equation (16) since $X_4 \neq (b_3^2)^{-1}(p_4)$ and, in equilibrium, $X_3 \neq (b_4^3)^{-1}(b_1^3(X_1))$. ∎

For this reason, to ensure consistency with the specified equilibrium bidding strategies, we limit the support of the underlying signals $(X_1, \ldots, X_N)$ to regions which — *at the estimated parameter values* θ — would yield the given dropout order. In other words, given parameter values θ^9, we limit the support of $(X_1, \ldots, X_N)$ to a region such that

$$b_i^{N-i}(X_i; \theta) = \min_{j=i,\ldots,N} b_j^{N-i}(X_j; \theta), \quad \text{for } i = 1, \ldots, N. \tag{18}$$

Recall our indexing convention, stated earlier, that bidder i drops out at the end of round $N - i$. For an N-bidder English auction, there will be $N(N-1)/2$ such constraints.[10]

These conditions induce a truncated distribution for the observed bids. Define the set $\mathcal{T}(\theta)$ as the set of draws from $F(X_1, \ldots, X_N \mid \theta$, observed bid ordering) which satisfy the consistency conditions (18), conditional on the values θ.

Then the joint density function of all the observed bids is given by:

$$g(p_0, \ldots, p_{N-2}) = \begin{cases} \dfrac{f * J}{P(\mathcal{A})} & \text{if } \vec{X} \in \mathcal{T}(\theta) \\ 0 & \text{otherwise} \end{cases} \tag{19}$$

---

[9] Unfortunately, as will be discussed below, these truncation conditions depend on θ, the estimated parameters of the F distribution. The resulting maximum likelihood estimation problem is "non-regular", and the estimates will not have a limiting normal distribution.

[10] For the log-normal model in Hong and Shum (1997), these constraints can be expressed as a set of inequalities which are linear in the observed dropout prices and nonlinear in the model parameters.

where, as before, $J = J(X_2, \ldots, X_N \mid p_0, \ldots, p_{N-2})$ is the Jacobian of the transformation from the private signals to the observed prices and $f = f(\phi_N^0(p_0; \theta), \ldots, \phi_2^{N-2}(p_{N-2}; \theta))$.[11]

Note that this consistency problem does not appear in symmetric models, where (cf. Milgrom and Weber (1982) pp. 1104-5) $b_i^j(x) = b_{i'}^j(x)$, for all bidders i, i' and rounds j. The symmetry assumption implies that, for all parameter values θ and across all rounds, the ordering of the signals will always be the same as the ordering of targeted dropout prices. Conditioning on the observed dropout order, then, is enough to ensure that, for all possible values of θ, any draw from $F_X(X_1, \ldots, X_N \mid \theta)$ will satisfy the consistency conditions (18).

For a different reason, consistency problems do not arise in single-round auction models *regardless of bidder asymmetries*. Even though conditioning on the observed bid ordering is not enough to ensure that draws from $F_X(X_1, \ldots, X_N \mid \theta)$ will satisfy the restrictions in (18) when bidder asymmetries are present, *any* ordering of the bids will be consistent with the specified equilibrium bidding strategies in single-round auctions. This is because in single-round auctions, unlike the ascending auction considered above, bidder i never learns the private signal of any other bidder, so that it is possible to observe a given bid independent of the realized bids for any of the other bidders. This is not the case in the ascending auction models in which a bidder's equilibrium bid depends on the realized bids for the bidders who have already dropped out.

4. Estimation strategies

4.1. MINIMUM DISTANCE (LEAST SQUARES) ESTIMATION

Once $g(p_1, \ldots, p_N; \theta)$, the joint density (with parameters θ) of the observed bids, has been derived, various estimators of θ are available. The first we consider is minimum distance (method of moments) estimation, which attempts to match sample moments of the observed bids against theoretical moments of the G distribution, computed at each parameter value.

Even under assumptions of symmetry and independence, the moments of G in first-price auction models may not be expressible analytically and are perhaps difficult to evaluate numerically. Furthermore, if the researcher only observes winning bid data, the moments of order statistics are even more difficult to calculate. Similarly, for ascending auction models which accommodate both asymmetries and common values, G is a multivariate

---

[11] Given the nonlinearity of the consistency conditions, the truncation probability $P(\mathcal{A})$ will likely require simulation methods to compute. See Hong and Shum (1997) for more details.

distribution, the moments of which are multivariate integrals which often are not expressible in closed form. Numerical integration techniques are inadequate once the dimension of G (i.e., the number of bidders in the auction) exceeds 4. For these reasons, we suggest adapting the simulated method of moments approach of Laffont, Ossard and Vuong (1995), in which the moments are approximated using Monte Carlo integration techniques.

We consider a least squares objective function, i.e., an estimator which minimizes the sum of squared deviations between the moments in the data and the theoretical sample moments of the G distribution:

$$(1/T)\sum_{t}\sum_{k=1}^{N_t}(p_k^t - E_g p_k^t)^2 \qquad (20)$$

where T is the number of auctions, N_t is the number of bidders in the tth auction, p_k^t is the kth bid in the tth auction, and the expectation is taken with respect to the G distribution, which perhaps does not exist in closed form.

The procedure for simulating $E_g p_k^t$ takes the following steps for draws $s = 1, \ldots, S$:

1. For each parameter vector θ that characterizes the joint distribution F, draw $X_1^s, \ldots, X_N^s$ from the marginal distribution F_X, holding the seed constant for random number generation across different values of θ.

2. Given the parameter value θ, evaluate the bids which correspond to the drawn signals: $p_1^s = b_1(X_1^s; \theta), \ldots, p_N^s = b_N(X_N^s; \theta)$.

3. (For asymmetric ascending auction only) If the bids $p_1^s, \ldots, p_N^s$ satisfy the consistency conditions (18), we retain this draw. Otherwise we discard this draw and repeat the above until we obtain a draw which satisfies these consistency conditions. This is the simplest type of "acceptance/rejection" method for sampling from a conditional distribution.[12]

Given S draws (or S accepted draws, for the case of the asymmetric ascending auction), we approximate the first moment of the bids as:

$$\text{Simulated } E_g p_k^t = (1/S)\sum\sum_s p_k^s, k = 1, \ldots, N_t. \qquad (21)$$

---

[12] More sophisticated sampling schemes, such as the sequential GHK simulator and Gibbs sampling, are described in Hajivassiliou and McFadden (1998).

Under standard conditions, this nonlinear least squares estimator is consistent and asymptotically normal, as S and N approach ∞.[13]

This approach is applicable to any auction model, provided we can simulate the moments of the G distribution, which require derivation of the equilibrium bid functions $b_i(\cdots ; F)$, $\forall i$. Note that explicit derivation of G, the equilibrium distribution of the bids, is not necessary for this estimation procedure, in contrast for maximum likelihood estimation. One main advantage of simulation techniques is the ease in simulating moments of an otherwise intractable (in this case, the G) distribution.

Given our distributional assumptions regarding the unobserved X's, we throw away information by only using the first moments for purposes of estimation. The distribution of the winning bid in an auction will be asymmetric, even assuming that the private values themselves are drawn from symmetric distributions. Therefore, in situations where only the winning bid is observed (as in Laffont, Ossard and Vuong (1995)), nonlinear regression which attempts only to match the observed winning bids to the *mean* of the winning bid distribution can be particularly inefficient.[14] Presumably, this problem would be less severe in situations where the researcher observes all of the bids from a given auction.

A special case deserves mention here. In their symmetric IPV framework, Laffont, Ossard and Vuong (1995) derive the conditional mean of the winning bid distribution in an interesting manner. They invoke the revenue equivalence theorem under which the expected revenue (i.e., winning bid) from a first- and second-price auction would be equivalent. Since they only observe the winning bid for their auctions, this theorem ensures that, in equilibrium, the winning bid will have the same expectation as the second-highest draw out of N draws from the H distribution (which is the winning bid in a symmetric IPV second-price auction), with corresponding density function given above in equation 5. In their simulated nonlinear least squares framework, they use this theorem to avoid having to simulate the equilibrium bid function (in equation 3) for any number of given draws of $(X_1, \ldots, X_N)$. However, this approach works only for the symmetry

---

[13] Laffont, Ossard and Vuong (1995) showed that, due to the linearity of their simulator in the draws, the simulated nonlinear least squares estimator is consistent even with a fixed number of simulated draws. However, the asymptotic variance of the estimator must be adjusted to take into account the variance introduced by the finite number of simulate draws. However, for the case of the asymmetric ascending auction with common value components, as in Hong and Shum (1997), the truncation probability also needs to be simulated for each vector of observed bids, and in the case the simulated moment becomes nonlinear in simulation draws. Therefore the simulated nonlinear least square estimator in this case is only consistent when the number of simulated draws increases with the sample size.

[14] Thanks to Samita Sareen for this insight.

IPV framework which they consider, and is not generalizable to alternative
auction paradigms.

4.2. MAXIMUM LIKELIHOOD ESTIMATION

Direct maximum likelihood estimation, on the other hand, utilizes all the
information embodied in the researcher's distributional assumptions.

However, for several auction models, it turns out that equilibrium behavior of the bidders implies that the support of the observed bids depends
parameters of the F distribution, which we are trying to estimate. Hong
(1998) shows that the resulting maximum likelihood estimates of these
parameters, while consistent, will not asymptotically normal.[15] Next, we
discuss several examples which have arisen in the literature.

4.2.1. First price auctions

For first-price auction models, equilibrium bidding behavior implies that
the support of the data depends on the parameters of the F distribution.
Assume that the (common) support of each X_i is $[\underline{x}, \overline{x}]$.

Both Laffont, Ossard and Vuong (1995) and Donald and Paarsch (1993)
note that, for the symmetric IPV first-price auction model, the upper bound
of the support for any bid typically depends on the parameters of the
H distribution. To see this, consider the equilibrium bid function for this
model in equation (3), which is reproduced here:

$$b(X_i) = X_i - \frac{1}{(H(X_i))^{N-1}} \int_{p_0}^{X_i} (H(X_i))^{N-1} dx. \tag{22}$$

Given that $b(X_i)$ is increasing in X_i, the upper bound of the support of any
observed bid is $b(\overline{x})$ — the bid that a bidder who observes a signal $\overline{x}$ would
submit. This will be a function of the parameters of the H distribution.[16]

Similar problems arise in asymmetric and non-independent first-price
auction models, such as that considered by Bajari (1996). The absence of

---

[15] Essentially, in these "nonregular" cases, the MLE is derived from a constrained
optimization problem, and is therefore not a root of the unconstrained maximum likelihood score function. In the "regular" case, the asymptotic normal distribution of the
MLE is derived by expanding this score function around the true parameter value. This
will not work in the nonregular case. See Newey and McFadden (1994, pp. 2141ff.) for
more details. In fact, the MLE is super-consistent, converging at rate T to a mixture of
exponential distributions.

In contrast, the simulated method of moments estimator suggested in the previous
section *is* a root of the first-order condition of the least squares objective function (20),
so that asymptotic normality obtains.

[16] An exception is where $\overline{x} = +\infty$, in which case $\lim_{x \to +\infty} b(x) = +\infty$ and the
regularity condition holds.

a clear asymptotic theory for the MLE in these multivariate models favors alternative estimation techniques, such as the minimum distance estimator described above.

4.2.2. *Open auctions*

Under the IPV assumption, no standard regularity conditions are violated in the second-price and ascending auctions models described earlier, because the equilibrium bid function is simply the identity function, and the support of p_i is therefore $[\underline{x}, \overline{x}]$, independently of θ.

However, these problems will crop up again in asymmetric models. As we pointed out earlier, the consistency restrictions (18) impose truncation conditions on the support of the bids observed in an asymmetric ascending auction which depend on θ. Unlike the constraints in first-price auction models discussed in the previous section, these constraints are multivariate (e.g., $l(p_1, \ldots, p_N; \theta) \geq 0$). Very little work has been done on the asymptotics of the MLE in these cases. In particular, although the maximum likelihood estimator is still consistent, its asymptotic distribution is unknown.

5. Conclusions

This paper illustrates in general terms the basic methodology of structural estimation using auction data. It provides a unified view of the common structure underlying structural econometric auction models under various model paradigms. We identify the most crucial steps in building a structural econometric model and discuss the estimation strategies for implementing these models.

The close dependence on a game-theoretic foundation is most the main advantage and disadvantage of structural auction models. While the economic theory provides an efficient framework for econometric estimation and allows for sharp prediction from the estimation results, a structural model is not robust to misspecification and to deviations between the assumptions in theoretical models and the rules of real-world auctions. Exactly the opposite can be said about reduced form approaches.

A compromise between the structural and the reduced form approaches would be to use very general behavioral assumptions — general enough to apply across a number of auction paradigms — in deriving the mapping between bidders' signals and their observed bids, thus retaining the flavor of structural modeling without relying fully on the equilibrium specifications of theoretical auction models. Recent work by Haile (1998) follows such an approach.

References

Bajari, P. (1996): "A Structural Econometric Model of the First Price Sealed Bid Auction: With Applications to Procurement of Highway Improvements", Mimeo, Stanford University.

Bulow, J., M. Huang and P. Klemperer (1996): "Toeholds and Takeovers" , Stanford Graduate School of Business working paper, # 1393.

Donald, S., and H. Paarsch (1993): "Piecewise Pseudo-Maximum Likelihood Estimation in Empirical Models of Auctions", *International Economic Review*, 34, 121-148.

Haile, P. (1998): "Structural Estimation of Oral Ascending Auctions", University of Wisconsin, Madison, mimeo.

Hajivassiliou, V., and D. McFadden (1998): "The Method of Simulated Scores for the Estimation of LDV Models", *Econometrica*, pp. 863-896.

Hajivassiliou, V., and P. Ruud (1994): "Classical Estimation Methods for LDV Models Using Simulation", in *Handbook of Econometrics, Vol. 4*, ed. by R. Engle and D. McFadden, North Holland.

Hendricks, K., and R. Porter (1988) "An Empirical Study of an Auction with Asymmetric Information", *International Economic Review*, 78, 865-883.

Hong, H. (1998): "Non-regular Maximum Likelihood Estimation in Auction, Job Search and Production Frontier Models", mimeo.

Hong, H., and M. Shum (1997): "The Econometrics of English Auctions", mimeo.

Laffont, J.J., H. Ossard and Q. Vuong (1995): "Econometrics of the First-Price Auctions", *Econometrica*.

Maskin, E., and J. Riley (1996): "Asymmetric Auctions", mimeo.

Milgrom, P., and R. Weber (1982): "A Theory of Auctions and Competitive Bidding", *Econometrica*, 50, 1089-1122.

Newey, W., and D. McFadden (1994): "Large Sample Estimation and Hypothesis Testing", in *Handbook of Econometrics, Vol. 4*, ed. by R. Engle and D. McFadden, North Holland.

Paarsch, H. (1991): "Deriving and Estimate of the Optimal Reserve Price: An Application to British Columbian Timber Sales", Univ. of British Columbia working paper, upcoming in *Journal of Econometrics*.

Paarsch, H. (1992): "Deciding between the common and private value paradigms in empirical models of auctions", *Journal of Econometrics*, 51, 191-215.

Vickrey, W. (1962): "Auctions and Bidding Games", in *Recent Advances in Game Theory*, Princeton University Press, reprinted in W. Vickrey, *Public Economics*, Cambridge University Press, 1996.

Vuong, Q., I. Perrigne, and E. Guerre (1996): "Optimal Nonparametric Estimation of First-Price Auctions", mimeo.

Vuong, Q., I. Perrigne, and T. Li (1997): "Auctions with Correlated Private Values", mimeo, University of Southern California.

Wilson, R. (1995): "Sequential Equilibria of Asymmetric Ascending Auctions", Forthcoming, *Economic Theory*.

CHAPTER 11

A MULTIPLICATIVE VARIANT OF THE SHAPLEY VALUE FOR FACTORIZING THE RISK OF DISEASE

MATTHIAS LAND
(matthias.land@ams.med.uni-goettingen.de)
and
OLAF GEFELLER
(gefeller@imbe.med.uni-erlangen.de)
Dept. of Medical Informatics, Biometry and Epidemiology
University of Erlangen–Nuremberg
Waldstraße 6, D-91054 Erlangen, Germany

Abstract. This article sets out to build the bridge between the mathematical philosophy of axiomatic approaches to cooperative game theory and practical concerns of statistical applications in epidemiology. It deals with the methodological task of appropriately assessing the impact of multiple characteristics of human behaviour, human constitution or environmental agents (the so called risk factors) on the disease load in the population. It is shown that this epidemiologic problem of multifactorial risk attribution can be formalized in a way that is comparable to the game–theoretic description of several players that act together in a grand coalition and are faced with the problem of dividing their profit fairly among them. In particular, epidemiologic methods of partitioning the risk of disease that are equivalent to the Shapley value are reviewed briefly. Moreover, a new parameter based on a multiplicative analogue of the Shapley value is introduced. Its application and the interpretation of its results are illustrated by epidemiologic data.

1. Introduction

Epidemiology is concerned with analyzing the occurrence patterns, the trends and the determinants of diseases in the population. This branch of medical science is based on an interdisciplinary approach involving among others empirical methods and knowledge from social and natural sciences and mathematical and applied statistics. In the midth of the twentieth century, several epidemiologic studies had far–reaching influence on public health decisions. The British Doctors Study, for example, was the first to establish an association between smoking and the occurrence of lung cancer (Doll & Hill, 1952). The increasing importance of epidemiologic results for medical research on the causes of disease and their implications for public health decisions was accompanied by a dynamic conceptual development of research strategies, study designs, measures of effects and statistical methods of "causal" inference. The latter particularly involved the refinement of multifactorial statistical models that realistically represent the complex interrelations of exposure to possible risk factors under study and the risk of developing the disease. This article focuses on statistical methods for quantifying proportions of the probability of disease that can be attributed to having been exposed to certain risk factors. It will be shown that these methods are closely related to both additive and multiplicative concepts of cooperative game theory.

The second section provides a brief review of fundamental concepts of risk attribution in epidemiology. In the third section it is shown that multifactorial approaches to risk attribution in epidemiology can be formalized in a way that is comparable to the game theoretic description of several players that act together in a grand coalition and are faced with the problem of dividing their profit fairly among them. In particular, epidemiologic methods of risk attribution that are equivalent to the Shapley value are reviewed briefly. In the fourth section a new method based on a multiplicative analogue of the Shapley value is introduced and its implications for epidemiologic problems of multifactorial risk attribution are discussed. Finally, the new method is illustrated by data from G.R.I.P.S., which is the Göttingen Risk-, Incidence- and Prevalence Study (Cremer et al., 1991).It involved male industrial workers, who were observed for myocardial infarction during a five year follow–up period. The occurrence of myocardial infarction was related to cholesterol levels and smoking habits. In G.R.I.P.S. multifactorial risk attribution provides an answer to the question:

> What proportion of the observed cases of myocardial infarction during the five year follow–up period can be attributed to smoking, to low HDL-cholesterol levels or to raised LDL- or VLDL-cholesterol levels, respectively, and how can one simultaneously account for the interrelations between the risk factors when assessing these proportions?

2. Epidemiologic principles of risk attribution

Before game theory–related solutions to the multifactorial risk attribution problem are derived this section briefly sketches the basic principles of risk assessment in epidemiology. More detailed introductions to methodological concerns of epidemiologic research, however, can be found in, among others, Rothman & Greenland (1998) or Kleinbaum et al. (1982).

2.1. RISK ATTRIBUTION IN THE CASE OF ONE EXPOSURE VARIABLE

In the simplest case the epidemiologic analysis focuses on a binary random variable D with $D = 1$ in case of a diseased subject and $D = 0$ otherwise. Furthermore, there is another categorical exposure variable E_1, the categories of which characterize the subject's level of exposure to the risk factor, where $E_1 = 0$ in case of the subject not being exposed.

The primary aim in epidemiologic studies is to analyze the relationship between D and E_1. The parameter preferred for assessing the strength of their association is the *relative risk* (RR). It is defined as the ratio of the risk of disease among the exposed to the risk of disease among the unexposed:

$$RR = \frac{P(D = 1|E_1 \neq 0)}{P(D = 1|E_1 = 0)}.$$

If the relative risk exceeds 1 the exposure is called a risk factor for the disease *on the individual level*. From a public health perspective, however, it is frequently more interesting to quantify the proportion of the observed disease events that can be attributed to having been exposed to the factor under study. Note that the relative risk parameter is not suitable for solving this problem of risk assessment *on the population level*. Suppose, for instance, that the exposed subpopulation is small in relation to the entire population. In that situation an exposure factor with a high relative risk may not account for a relevant proportion of the total number of disease events. On the other hand, a widespread exposure which is associated with a small relative risk may result in a substantial public health problem. Assessing the population impact of being exposed to a risk factor is thus based on both the relative risk of being exposed and the probability of exposure. The *attributable risk* parameter (AR) is used instead whenever such population–based questions have to be answered. It was introduced by Levin (1953) in order to quantify the proportion of lung cancer cases that were attributable to smoking. He defined the AR–parameter to be

$$AR = \frac{P(E_1 \neq 0) \cdot (RR - 1)}{P(E_1 \neq 0) \cdot (RR - 1) + 1}$$

$$= \frac{P(D = 1) - P(D = 1|E_1 = 0)}{P(D = 1)}. \tag{1}$$

The latter equation shows that the AR–parameter is equal to the proportionate reduction in the probability of disease when the population is prevented from being exposed to the risk factor, where it is assumed that elimination of the risk factor in the entire population leads to a new probability of disease, which is equal to the original probability among the unexposed.

2.2. RISK ATTRIBUTION IN THE CASE OF MULTIPLE EXPOSURE VARIABLES

Typically in observational epidemiologic studies multiple influential factors simultaneously affect the risk of disease. Therefore, throughout the article n categorical exposure variables $E_1, \ldots, E_n$ of primary interest are considered, where the ith variable has k_i categories and is equal to zero (the so called *low risk level*) in case of no exposure to the ith risk factor. Furthermore, it is common practice to include a permanent adjustment variable C with k_0 categories in order to account for different levels of exposure/disease associations in different subpopulations. The categories of C are supposed not to be affected by the elimination of primary interesting risk factors in the population. Age and sex, for example, are frequently used to define the variable C.

In this multifactorial situation epidemiologic risk analysis can be directed towards answering the following questions for each subset $S \subset \{E_1, \ldots, E_n\}$ of exposure variables:

> To what maximum extent can the probability of disease be lowered by completely reducing exposure to the risk factors described by variables in S to low risk levels? Or, to put it in other words, what share of the probability of disease is attributable to being exposed to at least one of the risk factors in the subset?

The attributable risk parameter that is suited to answering these questions is the *combined attributable risk*: for any subset $S \subset \{E_1, \ldots, E_n\}$ is defined as the relative difference between the original and the *reduced probability of disease* $P_S^{red}(D = 1)$ which hypothetically results from reducing exposure to the selected risk factors to low risk levels in the entire population:

$$AR(S) = \frac{P(D=1) - P_S^{red}(D=1)}{P(D=1)}.$$

The calculation of the reduced probability of disease is frequently based on the assumption that reducing all variables in S to low risk levels does *neither* affect the levels of the other exposure variables *nor* the levels of the permanent adjustment variable. Under this condition the reduced probability of disease is a weighted sum of conditional probabilities of disease, where the condition fixes the interesting variables at zero level and states that the other variables remain unchanged (Walter, 1980; Benichou, 1991):

$$P_S^{red}(D = 1) = \sum_{k=1}^{k_S} P(K = k) \cdot P\big(D = 1\big| \bigcap_{E_j \in S} \{E_j = 0\} \cap \{K = k\}\big), \quad (2)$$

where K is a categorical random variable with $k_S = k_0 \cdot \prod_{E_i \notin S} k_i$ categories, specifying all combinations of categories of C and $E_i \notin S$. The combined attributable risk resulting from this conception of the reduced probability of disease is adjusted for C and $E_j \notin S$.

3. The bridge between cooperative game theory and epidemiology

The combined (adjusted) attributable risk $AR(\{E_1, \ldots, E_n\})$ of all variables of primary interest quantifies the maximum percentage reduction in the disease rate when the population under study is prevented from being exposed to the interesting risk factors. In various epidemiologic situations there is an urgent need for methodological answers to the question of how much of this percentage is contributed by *single* exposures (Lagakos & Mosteller, 1986; Christoffel & Teret, 1991). Suppose, for instance, a uranium miner who smokes cigarettes and finally developed lung cancer was exposed to some amount of occupational radiation. In the case of the miner taking legal actions in order to fight for compensation a juridical problem arises: both exposures are known to be strong risk factors for lung cancer, but the employer is only responsible for radiation whereas the worker himself is responsible for smoking. So, how can the compensation for him be determined?

The solution to the general problem is inspired by an idea by Cox (1985) who pointed to the formal equivalence of multifactorial risk assessment in epidemiology and the mathematical formalism in cooperative game theory. Each variable can be compared to a player and any subset of variables specifying risk factors to be eliminated in the population corresponds to a coalition of players. The solution to the problem of multifactorial risk attribution can be found by interpreting the function $AR : \{S|S \subset \{E_1, \ldots, E_n\}\} \to \mathbb{R}$ mapping each subset of variables to their combined (adjusted) attributable risk as a cooperative n–person game in characteristic function form (Land & Gefeller, 1997). Note that an AR–function satisfies $AR(\emptyset) = 0$. Therefore, it is element of the system $\mathfrak{R}_n = \{R : \{S|S \subset \{E_1, \ldots, E_n\}\} \to \mathbb{R}|R(\emptyset) = 0\}$ of all epidemiologic risk functions, given n exposure variables of primary interest. The latter is isomorphic to the space $\mathfrak{G}_n$ of all games in characteristic function form with a fixed number of n players. Finally, applying the Shapley value $\Psi : \mathfrak{G}_n \to \mathbb{R}^n$ to an (adjusted) attributable risk function AR results in a vector of so called *partial attributable risks* (Eide & Gefeller, 1995):

$$
\begin{aligned}
\text{PAR}_i &= \Psi_i(AR) \\
&= \sum_{S \subset \{E_1,\dots,E_n\}: E_i \notin S} \frac{|S|! \, (n - |S| - 1)!}{n!} \, \delta_{E_i,S}(AR) \, , \quad i = 1,\dots,n,
\end{aligned}
$$

where for $R \in \mathfrak{R}_n$ and $S \subset \{E_1,\dots,E_n\}$ with $E_i \notin S$

$$
\delta_{E_i,S}(R) = R(S \cup \{E_i\}) - R(S).
$$

The components $\text{PAR}_1,\dots,\text{PAR}_n$ quantify shares of the probability of disease that can be attributed to having been exposed to the respective risk factors. Note, however, that any attributable risk function depends on the basic probability distribution P induced by the complete cross-classification of disease-, exposure- and subpopulation-variables. As a consequence of P being unknown the partial attributable risk itself is an unknown population characteristic, a parameter function the components of which have to be estimated from epidemiologic data. This estimation process is exemplified by the G.R.I.P.S. data in the fifth section.

Shapley introduced a set of axioms that uniquely determine the Shapley value. His additivity axiom, however, cannot be interpreted epidemiologically. Recently, a set of axioms, originally given by Young (1985), was transferred to the epidemiologic context (Land & Gefeller, 1997). His monotonicity, symmetry and efficiency axioms can be interpreted very well in epidemiologic applications (Gefeller et al., 1998).

Apart from applying the game theory related PAR–parameter, the problem of multifactorial risk attribution in epidemiology can alternatively be solved by leaving the additive pathway and transferring the Shapley value to a multiplicative context. This approach is outlined in the following section.

4. The multiplicative approach

The multiplicative approach to the problem of assessing the relevance of single exposure factors in the multifactorial setting focuses on the reduced probabilities of disease in proportion to the original probability of disease. For any subset $S \subset \{E_1,\dots,E_n\}$ the fraction

$$
CAR(S) = \frac{P_S^{red}(D = 1)}{P(D = 1)} = 1 - AR(S) \tag{3}
$$

is termed the combined *complementary attributable risk* of the variables in S. This proportion measures the hypothetical relative effect of reducing exposure to the risk factors described by S to low risk levels. The function that maps each subset S to $CAR(S)$ is termed *complementary attributable risk function*. A practical example of such a function estimated from the G.R.I.P.S. data can be found in the fifth section.

The hypothetically reduced probability of disease after simultaneously eliminating all n risk factors in the population results from multiplying the original probability of disease with $CAR(\{E_1, \ldots, E_n\})$. The latter is thus a measure for the combined public health relevance of all considered factors. For the purpose of assessing the relevance of the individual risk factors, however, it is interesting to find n partial proportions $\text{FCAR}_1, \ldots, \text{FCAR}_n$, called factorial complementary attributable risks, the product of which relates the completely reduced probability of disease to the actual probability of disease (see figure 1).

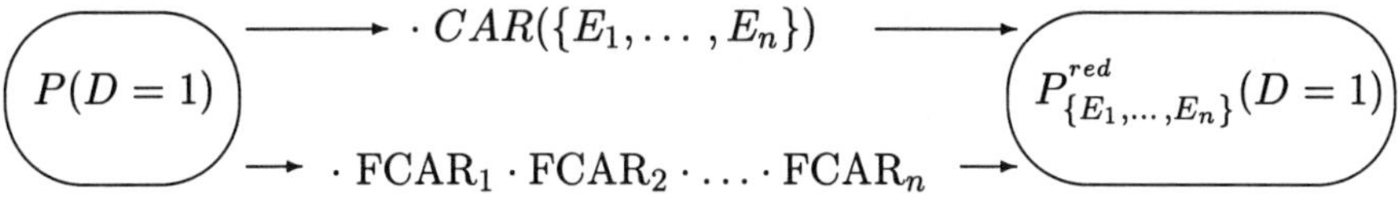

$$
\boxed{P(D = 1)}
\begin{array}{c}
\longrightarrow \cdot\, CAR(\{E_1, \ldots, E_n\}) \longrightarrow \\
\\
\longrightarrow \cdot\, \text{FCAR}_1 \cdot \text{FCAR}_2 \cdot \ldots \cdot \text{FCAR}_n \longrightarrow
\end{array}
\boxed{P^{red}_{\{E_1, \ldots, E_n\}}(D = 1)}
$$

Figure 1.   The principle of factorizing the combined complementary attributable risk of all variables for assessing the public health relevance of the individual risk factors.

In section 4.3 the FCAR_i–parameter is defined to quantify the "expected" reduced probability of disease that results from eliminating exposure to the ith risk factor in relation to the original probability of disease. Consequently, a small FCAR_i component indicates a risk factor with a high public health importance. The estimation of the FCAR is based on a new multiplicative variant of the Shapley value which is defined in the following.

4.1. DEFINITIONS AND NOTATION

A *positive risk function* is any function R that maps the system of all subsets of n exposure variables of primary interest into $I\!\!R^*_+$, the set of positive real numbers, and satisfies $R(\emptyset) = 1$. The multiplicative group of positive risk functions is denoted by $\mathfrak{R}^*_n$. Note that the positivity of the risk function is a need for the existence of the multiplicative Shapley value defined in section 4.2.

An n–dimensional *factorial risk assessment functional* is any mapping $F : \mathfrak{R}^*_n \to (I\!\!R^*_+)^n$ that satisfies the so called factorizing condition

$$
R(\{E_1, \ldots, E_n\}) = \prod_{i=1}^{n} F_i(R) \qquad \text{for all } R \in \mathfrak{R}^*_n.
$$

The solution to the problem of finding multiplicative components that quantify the relative benefit of reducing the respective factors to low risk levels can thus be solved by applying a factorial n–dimensional risk assessment functional to a complementary attributable risk function.

In most epidemiologic applications we are particularly interested in symmetric risk assessment functionals. This feature is defined analogously to the symmetry of values, which means that $F : \mathfrak{R}_n^* \to (I\!\!R_+^*)^n$ is called symmetric if

$$F_{\pi(i)}(R_\pi) = F_i(R) \tag{4}$$

for all permutations π of the integers $1, \ldots, n$, for all $i \in \{1, \ldots, n\}$, and for all $R \in \mathfrak{R}_n^*$. In (4) the values $R_\pi(S)$ of the permutated risk function are equal to $R(\{E_i | E_{\pi(i)} \in S\})$ for all $S \subset \{E_1, \ldots, E_n\}$.

A second important feature focuses on the relative contributions of single factors to the values of a positive risk function. For $i \in \{1, \ldots, n\}$ and $R \in \mathfrak{R}_n^*$ they are defined as

$$\rho_{E_i,S}(R) = \frac{R(S \cup \{E_i\})}{R(S)} \quad \text{for all } S \subset \{E_1, \ldots, E_n\} \setminus \{E_i\}.$$

A risk assessment functional $F : \mathfrak{R}_n^* \to (I\!\!R_+^*)^n$ satisfies *marginal rationality* if with regard to all positive risk functions $R_a \in \mathfrak{R}_n^*$ and $R_b \in \mathfrak{R}_n^*$ as well as any exposure variable $E_i \in \{E_1, \ldots, E_n\}$ the following implication is valid:

$$\rho_{E_i,S}(R_a) \leq \rho_{E_i,S}(R_b) \quad \forall S \subset \{E_1, \ldots, E_n\} \setminus \{E_i\}$$
$$\Rightarrow F_i(R_a) \leq F_i(R_b).$$

Note that marginal rationality is the multiplicative counterpart of Young's strong monotonicity (Young, 1985). A marginally rational risk assessment functional $F : \mathfrak{R}_n^* \to (I\!\!R_+^*)^n$ thus satisfies a multiplicative version of the marginality principle which means that for each positive risk function R the ith component $F_i(R)$ depends only on the relative contributions of the corresponding risk factor.

4.2. A MULTIPLICATIVE VARIANT OF THE SHAPLEY VALUE

Symmetry and marginal rationality are desirable properties of factorial risk assessment functionals. The following result, however, expresses the fact that there is one and only one n–dimensional factorial risk assessment functional that simultaneously satisfies both conditions.

THEOREM 1. *The functional* $\Phi : \mathfrak{R}_n^* \to (I\!\!R_+^*)^n$ *assigning to every positive risk function* $R \in \mathfrak{R}_n^*$ *the values*

$$\Phi_i(R) = \prod_{S \subset \{E_1,\dots,E_n\}\backslash\{E_i\}} \left(\rho_{E_i,S}(R)\right)^{\frac{|S|!\cdot(n-|S|-1)!}{n!}}, \quad i = 1,\dots,n, \quad (5)$$

is symmetric and marginally rational. Furthermore, Φ *is the only n-dimensional factorial risk assessment functional that simultaneously fulfills these conditions.*

The components of Φ are weighted geometric means of relative contributions to positive risk functions. The functional Φ is thus the multiplicative variant of the Shapley value, which can be seen by analyzing figure 2. Note that the multiplicative group $\mathfrak{R}_n^*$ of positive risk functions is isomorphic to the additive group $\mathfrak{R}_n$ of general epidemiologic risk functions (or to the space $\mathfrak{G}_n$ of n–person games, respectively) via a multidimensional logarithm. The same is obviously true for the corresponding n–dimensional real groups. The diagram below is commutative and therefore Φ is the multiplicative functional induced by the Shapley value which means that $\Phi(R)$ is equal to $\exp\big(\Psi(\ln \circ R)\big)$ for each positive risk function R.

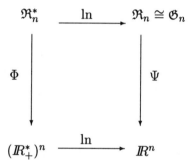

Figure 2. The functional Φ is the multiplicative functional induced by the Shapley value Ψ.

The above theorem is the multiplicative counterpart to the uniqueness statement proved by Young (1985). Consequently, uniqueness of Φ can be proved by transferring Young's proof from the additive to the multiplicative context according to these isomorphisms.

An interpretation of Φ can be derived from the well known fact that for each risk function $R \in \mathfrak{R}_n$ and for each variable $E_i \in \{E_1,\dots,E_n\}$ the Shapley value $\Psi_i(R)$ can be represented as

$$\Psi_i(R) = I\!\!E\big(\delta_{E_i,\mathfrak{S}}(R)\big),$$

where $\mathfrak{S}$ is a random variable whose values are subsets of $\{E_1,\dots,E_n\} \setminus \{E_i\}$ and $I\!E$ denotes the linear expectation with respect to the probability density P' on the subsets of $\{E_1,\dots,E_n\} \setminus \{E_i\}$ that assigns identical probabilities to the random subsets of the same size and assigns the same weight to each possible cardinality:

$$P'(\mathfrak{S}=S) = \frac{1}{n \cdot \binom{n-1}{|S|}} \quad \text{for each } S \subset \{E_1,\dots,E_n\} \setminus \{E_i\}.$$

Now let $I\!E^*$ denote the geometric expectation with respect to P' (i.e. $I\!E^* \equiv \exp \circ I\!E \circ \ln$). Note that $I\!E^*$ is related to $I\!E$ in the same way as the geometric mean is related to the arithmetic mean. Therefore, $I\!E^*$ is the multiplicative variant of the linear expectation allowing the following representation of Φ. For each positive risk function R and for each variable E_i:

$$\begin{aligned}
\Phi_i(R) &= \exp\left(\Psi_i(\ln \circ R)\right) \\
&= \exp\left(I\!E\left(\delta_{E_i,\mathfrak{S}}(\ln \circ R)\right)\right) \\
&= \exp\left(I\!E\left(\ln(\rho_{E_i,\mathfrak{S}}(R))\right)\right) \\
&= I\!E^*\left(\rho_{E_i,\mathfrak{S}}(R)\right).
\end{aligned} \tag{6}$$

The ith component of the multiplicative variant of the Shapley value is thus the expected (multiplicative) contribution of the ith factor to the values of the positive risk function. Further mathematical characteristics of multiplicative values are discussed by Land (1999).

4.3. THE FCAR–PARAMETER

The epidemiologic question of how to solve the multifactorial risk attribution problem multiplicatively can finally be answered by defining the FCAR–parameter to result from applying the functional Φ to the CAR–function, whenever this function is positive, which means that for each probability distribution P on the entirely cross–classified system of disease, exposure and adjustment variables the FCAR is defined with respect to the CAR–function calculated from formula (3) as

$$FCAR = \Phi(CAR), \tag{7}$$

whenever $CAR \in \mathfrak{R}_+^*$. Conditions on P that guarantee the positivity of the CAR–function are discussed later.

The concept of factorial complementary attributable risk has been defined for exposure variables that are allowed to have multiple high risk categories. The total number of categories has an effect on the calculation of the CAR–function: increasing the number k_i of categories for variable E_i increases the values of k_S, $S \subset \{E_1, \ldots, E_n\} \setminus \{E_i\}$ in formula (2). Therefore, the number of categories of E_i influences *all* components FCAR$_1, \ldots$, FCAR$_n$.

As a consequence of formula (6), FCAR$_i$ quantifies the expected reduced probability of disease after (additionally) eliminating *all* high risk categories of the ith risk factor in the population in relation to the probability of disease that results from not changing the levels of exposure to this factor in the population:

$$\text{FCAR}_i = I\!\!E^* \left(\frac{P^{red}_{\mathfrak{S} \cup \{E_i\}}(D = 1)}{P^{red}_{\mathfrak{S}}(D = 1)} \right), \quad \text{for } i = 1, \ldots, n. \quad (8)$$

The FCAR has smaller values for a risk factor with a higher public health importance. In order to avoid this reciprocity it is frequently more informative to base the analysis on the so called *factorial attributable risk* FAR with FAR$_i = 1 - \text{FCAR}_i$ for each variable $E_i \in \{E_1, \ldots, E_n\}$. This parameter can be interpreted very well because for each $E_i \in \{E_1, \ldots, E_n\}$ equation (6) implies:

$$\text{FAR}_i = \frac{I\!\!E^* \left(P^{red}_{\mathfrak{S}}(D = 1) \right) - I\!\!E^* \left(P^{red}_{\mathfrak{S}}(D = 1 | E_i = 0) \right)}{I\!\!E^* \left(P^{red}_{\mathfrak{S}}(D = 1) \right)}. \quad (9)$$

When comparing equations (1) and (9) it is obvious that the components of the FAR–parameter can be interpreted as attributable risks, where the actual probabilities of disease are exchanged by the expected probabilities of disease.

When using the FCAR or the FAR in real epidemiologic situations the parameters need to be justified. This could be done by referring to the uniqueness of Φ and arguing that there is no alternative functional satisfying the reasonable conditions of symmetry and marginal rationality. The latter condition was motivated in the context of abstract positive risk functions, but what does it mean when it is applied to complementary attributable risk functions? Are the conditions of symmetry and marginal rationality strong enough to imply uniqueness of the FCAR–parameter when they are restricted to the subclass of CAR–functions? The last question, however, can be answered under the *non–triviality condition*. It states that — conditional on having developed the disease — a subject is exposed to each combination of levels of the risk factors and the adjustment variable with positive probability:

$$P\left(\bigcap_{i=1}^{n}\{E_i = e_i\} \cap \{C = c\} \,\middle|\, D = 1\right) > 0, \tag{10}$$

for all possible values e_1 of $E_1, \ldots, e_n$ of E_n and c of C.

This condition is fulfilled in most epidemiologic situations and it implies that each complementary attributable risk function is a positive risk function. But the reverse is also true: it can be proved that for each positive risk function R there exists a probability distribution P on the entirely cross–classified system of exposure and disease levels satisfying condition (10) and inducing a CAR–function that equals R. Therefore, the group of positive risk functions is equal to the system of all CAR–functions and the condition of marginal rationality as well as the uniqueness result for Φ can be reformulated in terms of these functions. This fact implies a uniqueness result for the FCAR–parameter in purely epidemiologic terms which is outlined in the following.

Suppose that the effects of reducing exposure to the ith risk factor to the low risk level has to be assessed for two subpopulations A and B (the subpopulations of men and women, for example) separately. Two probability distributions P_A and P_B are necessary to understand and describe the association of disease and exposure to the risk factors in both subpopulations. Furthermore, suppose that eliminating the ith risk factor leads to a relative decrease in the risk of disease in population A that is at least as substantial as the relative decrease in the risk of disease in population B, and suppose that this is valid independently of the set of risk factors eliminated before. Formally this means that

$$\rho_{E_i,S}(CAR_A) \leq \rho_{E_i,S}(CAR_B) \quad \text{for all } S \subset \{E_1, \ldots, E_n\} \setminus \{E_i\},$$

where CAR_A and CAR_B are the complementary attributable risk functions calculated with respect to P_A and P_B, respectively. In that situation the public health importance of the ith risk factor is at least as high in population A as in population B. It is reassuring that the FCAR of the ith factor will definitely represent this "ranking" among both subpopulations, which means that the FCAR$_i$ calculated from P_A is at least as small as the FCAR$_i$ calculated from P_B. This characteristic of the FCAR–parameter is very important in epidemiologic applications because it ensures the rational comparability of different subpopulations, and using the parameter for assessing the public health relevance of single risk factors can be justified by the fact that no multiplicative risk parameter apart from the FCAR satisfies this epidemiologic condition and is essentially independent from the enumeration of variables.

In the next section the calculation and interpretation of the FCAR–parameter is illustrated by data from the G.R.I.P. study.

5. G.R.I.P.S.: an epidemiologic example

The Göttingen risk incidence and prevalence study (G.R.I.P.S.) involved 6029 male industrial workers aged 40-60 years, who were observed for myocardial infarction during a five year follow-up period. The binary disease variable D is thus equal to 1 in case of the subject having suffered from myocardial infarction during this period. The occurrence of infarction was related to cholesterol levels and smoking habits so that four binary exposure variables T, L, H and V of primary interest are defined as follows:

- $T = 0$ for nonsmokers, $T = 1$ otherwise,
- $L = 0$ for LDL-cholesterol < 160 mg/dl, $L = 1$ otherwise,
- $H = 0$ for HDL-cholesterol > 35 mg/dl, $H = 1$ otherwise,
- $V = 0$ for VLDL-cholesterol < 30 mg/dl, $V = 1$ otherwise.

Calculations involve an additional adjustment variable C of minor interest. It has multiple categories characterizing the subject's age, family predisposition, alcohol consumption, blood pressure and glucose level. The CAR–function was estimated from the G.R.I.P.S. data via a logistic regression approach (Greenland & Drescher, 1993). For details see Muche (1995). Calculations were based on equation (3) and the results can be found in table I.

The estimated FCARs for the blood lipoprotein fractions and smoking in the G.R.I.P.-study displayed in table I quantify fractions of the relative change in the risk of disease that can be attributed to the separate risk factors when eliminating all of them in the population. As a consequence of equation (9) the estimated FAR–components in the same table can be interpreted as shares of the expected probability of disease that can be attributed to the ith risk factor. Therefore, the estimated components of both the FCAR and the FAR can be interpreted as measures for the relative preventive potentials of intervention campaigns targeting the corresponding risk factors. It can be seen that eliminating high risk levels of LDL cholesterol in the population is hypothetically associated with the most substantial relative decrease in the risk of infarction.

The same ranking of exposures can be observed when the partial attributable risks are calculated:

$$\widehat{\text{PAR}}_T = 0.2111, \quad \widehat{\text{PAR}}_L = 0.4048, \quad \widehat{\text{PAR}}_H = 0.0796, \quad \widehat{\text{PAR}}_V = 0.0742.$$

This imposes the question whether the conceptual differences between the multiplicative and the additive view are mainly of game theoretic interest

TABLE I. The attributable (2nd column) and complementary attributable risks (3rd column) as well as the relative contributions of T, L, H and V (4th – 7th column) and their factorial complementary attributable risks (last line) estimated from the G.R.I.P.S. data.

S	$\widehat{AR}(S)$	$\widehat{CAR}(S)$	$\rho_{T,S}(\widehat{CAR})$	$\rho_{L,S}(\widehat{CAR})$	$\rho_{H,S}(\widehat{CAR})$	$\rho_{V,S}(\widehat{CAR})$
$\emptyset$	0	1	0.6303	0.4227	0.8278	0.8331
$\{T\}$	0.3697	0.6303		0.4587	0.8916	0.8862
$\{L\}$	0.5773	0.4227	0.6839		0.8524	0.8819
$\{H\}$	0.1722	0.8278	0.6789	0.4353		0.8916
$\{V\}$	0.1669	0.8331	0.6705	0.4475	0.8860	
$\{T,L\}$	0.7109	0.2891			0.8661	0.8883
$\{T,H\}$	0.4380	0.5620		0.4456		0.9043
$\{T,V\}$	0.4414	0.5586		0.4597	0.9098	
$\{L,H\}$	0.6397	0.3603	0.6950			0.9117
$\{L,V\}$	0.6272	0.3728	0.6888		0.8812	
$\{H,V\}$	0.2619	0.7381	0.6885	0.4451		
$\{T,L,H\}$	0.7496	0.2504				0.9193
$\{T,L,V\}$	0.7432	0.2568			0.8962	
$\{T,H,V\}$	0.4918	0.5082		0.4530		
$\{L,H,V\}$	0.6715	0.3285	0.7008			
$\{T,L,H,V\}$	0.7698	0.2302				

	Factorial Complementary Attributable Risks			
	$\widehat{FCAR}_T$	$\widehat{FCAR}_L$	$\widehat{FCAR}_H$	$\widehat{FCAR}_V$
	0.6743	0.4430	0.8711	0.8845

	Factorial Attributable Risks			
	$\widehat{FAR}_T$	$\widehat{FAR}_L$	$\widehat{FAR}_H$	$\widehat{FAR}_V$
	0.3257	0.5570	0.1289	0.1155

or whether these differences represent two independent informative aspects of multifactorial risk attribution in epidemiology. It takes further research to answer this question in appropriate generality. With respect to the G.R.I.P.S. data, however, it can be observed that the estimated relative contributions displayed in table I are nearly independent from the selection of the reference set $S \subset \{E_1, \ldots, E_n\}$. This is not true for the corresponding additive contributions which range from 0.0983 to 0.3697 for variable T, from 0.2780 to 0.5773 for variable L, from 0.0266 to 0.1722 for variable H and from 0.0202 to 0.1669 for variable V. Or, to put it in game theoretic words, the estimated CAR–function itself is nearly a multiplicative set function, whereas $\widehat{AR}$ is far from being additive. But this fact has an interesting consequence within an epidemiologic context: exposure variables whose additive or multiplicative relative contributions vary in a small range can be compared to each other more reasonably than variables with relative contributions varying in wide and possibly overlapping ranges. It may thus be argued that the multiplicative view is more appropriate for modelling the multifactorial situation in G.R.I.P.S. However, the nearly multiplicative structure of the estimated CAR-function might be related with the logistic regression model involved in the estimation procedure. This consideration points to the need for further research concerning the possible effects of the estimation procedure on the FCAR or the multiplicative Shapley value, respectively.

Acknowledgements

This work was supported by a grant from the Deutsche Forschungsgemeinschaft (grant no. Ge 637/3-2).

References

Benichou, J., 1991: Methods of adjustment for estimating the attributable risk in case–control studies: a review. *Stat. Med.*, **10**, 1753–1773.

Christoffel, T. and Teret, S. P., 1991: Epidemiology and the law: courts and confidence intervals. *Amer. J. Public Health*, **81**, 1661–1666.

Cox, L. A. Jr., 1985: A new measure of attributable risk for public health applications. *Man. Sci.*, **31**, 800–813.

Cremer, P., Nagel, D., Labrot, B., Muche, R., Elster, H., Mann, H. and Seidel, D., 1991: *Göttinger Risiko-, Inzidenz- und Prävalenzstudie (GRIPS)*. Springer–Verlag, Heidelberg.

Doll, R. and Hill, A. B., 1952: A study of the aetiology of carcinoma of the lung. *Br. Med. J.*, **2**, 1271–1286.

Eide, G. E. and Gefeller, O., 1995: Sequential and average attributable fractions as aids in the selection of preventive strategies. *J. Clin. Epidemiol.*, **48**, 645–655.

Gefeller, O., Land, M. and Eide, G.E., 1998: Averaging attributable fractions in the multifactorial situation: assumptions and interpretation. *J. Clin. Epidemiol.*, **51**, 437-441.

Greenland, S. and Drescher, K., 1993: Maximum likelihood estimation of the attributable fraction from logistic models. *Biometrics*, **49**, 865–872.

Kleinbaum, D. G., Kupper, L. L. and Morgenstern, H., 1982: *Epidemiologic Research*. Lifetime Learning Publications, Belmont, California.

Lagakos, S. W. and Mosteller, F., 1986: Assigned shares in compensation for radiation–related cancers. *Risk Analysis* **6**, 345–357.

Land, M. and Gefeller, O. , 1997: A game–theoretic approach to partitioning attributable risks in epidemiology. *Biom J.* , **39**, 777–792.

Land, M., 1999: Multiplicative values for factorizing risks (in preparation).

Levin, M.L., 1953: The occurrence of lung cancer in man. *Acta Unio Internationalis Contra Cancrum*, **9**, 531–541.

Muche, R., 1995: *Variablenselektion in Kohortenstudien. Lipoproteine als Risikofaktoren für den Myokardinfarkt in der G.R.I.P-Studie.* Dissertation, Faculty of Medicine, University of Ulm, Ulm.

Rothman, K. J. and Greenland, S., 1998: *Modern Epidemiology.* Lippincott–Raven, Philadelphia.

Shapley, L. S., 1953: A value for n–person games. In: Kuhn, H. and Tucker, A. (Eds.): *Contributions to the theory of games II.* Ann. Math. Studies, **28**, 307–317.

Walter, S. D., 1980: Prevention for multifactorial disease. *Amer. J. Epidemiol.*, **112**, 409–416.

Young, H. P., 1985: Monotonic solutions of cooperative games. *Int. J. Game Theory*, **14**, 65–72.

CHAPTER 12

EXPERIMENTS ON AUCTIONS WITH RANDOM PRIZES AND EU/NON-EU BIDDERS*

LUCIA PARISIO (lucia.parisio@unimi.it)
Institute of Economics and Statistics
University of Milan, Milan, Italy

Abstract. Recent developments in the theory of auctions with random prizes played by non-EU bidders provide a series of behavioral prescriptions which can be investigated with experimental methods. In this paper, we present results of an experiment involving a large number of participants, held at the Centre for Experimental Economics (University of York, UK). The test of the theory is conducted by means of auction (English and second price sealed bid) data coupled with some outside the auction evidence of bidders' behavior towards risk. The latter evidence is obtained thanks to Neilson's (1992) "modified" version of the standard Becker et al. methodology. Experimental results indicate that bidders' behaviour is sensitive to the environment which is used to solicit their certainty equivalent of lotteries. It is also found that individual violations of Expected Utility are smoothed down at the price level, since our experimental data confirms revenue equivalence.

1. Introduction

Revenue equivalence is an important theoretical result in the theory of auctions based on the assumptions of risk neutral bidders who have non-random private and independent valuations of the item at stake[1]. In the

---

* With the usual disclaimers, I wish to thank Prof. John Hey and two anonymous Referees for helpful comments and suggestions.
[1] A survey of the main results in auction theory is contained into Milgrom and Weber (1982), McAfee and McMillan (1987) and Wolfstetter (1996).

160 PARISIO

English[2] and SPSB[3] [4] procedures revenue equivalence is obtained through a unique non-cooperative dominant strategy equilibrium where each bidder reveals his private non-random valuation of the object. Recent theoretical literature considers strategic equilibria for auction games in which the assumption of deterministic valuations is abandoned. Theoretical papers like Neilson (1994) and Karni and Safra (1986 and 1989) analyse English and SPSB procedures in which agents bid for random lotteries with independent private outcomes. The main result of the analysis is that Revenue Equivalence can be extended to the new random-prize setting only if bidders behave as expected utility maximisers (EUM). The revenue equivalence between English and SPSB auctions fails when bidders preferences do not obey the independence axiom of expected utility. In this case, it is shown that the revenue ranking of the auction procedures depends upon the way bidders' attitudes towards risk change in response to shifts towards more preferred lotteries.

The main results obtained by this literature provide the theoretical basis of the experimental work described in this paper.

Previous experiments based on auction theory with deterministic prizes devoted much attention to the revenue equivalence result[5]. The development of the theoretical literature of auctions with random payoffs opens a new field of experimental research. In the experiment that motivated this paper we first analyse revenue equivalence in English and SPSB auctions for independent private value lotteries. We then try and interpret our auction data on the basis of the theoretical relationship among bidders' preferences and bidding behaviour. To this end, a new methodology suggested by Neilson (1992) is used in our experiment to provide some outside-the-auction evidence of bidders' preferences over risky prospects. The latter evidence allows one to evaluate whether bidders' preferences alone explain auction behaviour or, alternatively, to what extent bidding competition influences agents' behaviour for given preferences.

Experimental works related to our paper evaluate the predictive power of expected utility theory at individual (non-market) and aggregate (market) level. Evans (1997) for example, found that expected utility predicts better in a market environment than at individual level. However, she notice

---

[2] In the English (Oral Ascending Bid) auction the auctioneer continuously raises the price until the last but one bidder exits the procedure. The winner is the last remaining bidder who pays the last announced price, i.e. the price at which the second highest bidder abandoned the competition.

[3] We shall use the following abbreviations: EU – expected utility; MBDM – modified Becker, De Groot, Marschak mechanism; SPSB – second price sealed bid

[4] In the SPSB auction bidders submit sealed bids and the winner, who is the highest bidder, pays a price equal to the second highest bid.

[5] Kagel (1995) presents a wide survey of this literature.

that the result may be influenced by the price rule selected[6]. Following the same line of research, we show that agents' behaviour significantly departs from the expected utility prescriptions at the individual level while such departures are smoothed down at the market level. In fact, prices formed in the English procedure are shown to be not significantly different from their SPSB auction counterparts, which in turn implies the expected revenue equivalence prescribed by EUT.

The next session contains some fundamental results in the theory of auctions without expected utility and highlights the implied main experimental hypotheses. The experimental design is described in Section 3 whereas section 4 contains comments on the results obtained from an experiment conducted at the Center for Experimental Economics (EXEC) of York University. Final comments are contained in section 5.

2. Auctions without expected utility

Let us consider n bidders who compete in two alternative auction procedures, namely English and SPSB. Assume that the auction prize takes the form of a random variable $X(r)$, having known distribution $F(.,r)$, where r represents a shift parameter designed in a way that higher values of r imply higher-valued prizes. Independent private values may be induced when r is a bidder-specific parameter drawn from a commonly known distribution G having support $[a, b]$. We assume first that bidders are expected utility maximisers with expected utility $u(\cdot)$, initial wealth equal to zero and $u(0) = 0, \forall u$.

In this standard context, Karni and Safra (1989) showed that both the English and the SPSB auctions have a dominant strategy $B(r_i)$ in which each agent i bids up to the certainty equivalent, of his random prize. In particular, the assumption of EUM bidders is a necessary and sufficient condition for revenue equivalence[7]. Therefore, for a given draw r_i, the optimal bid $B_i^* = B(r_i)$, solves:

$$\int_{X_a}^{X_b} u(X - B_i^*) \, dF(X, r_i) = 0 \qquad (1)$$

where $[X_a, X_b]$ indicates the support of $F(.,r)$.

The equivalence result does not longer hold when bidders do not satisfy the independence axiom of expected utility. Combining the results of Karni

---

[6] Evans (1997) implemented a fifth price auction (market setting) against a standard BDM mechanism (individual setting).

[7] The equivalence of English and SPSB auction procedures under EU is proved combining Karni and Safra (1989) Theorems 2 and 4.

e Safra (1989) Theorems 3 and 4, we obtain that: i) the revelation property
of ascending-bid auctions is maintained in a random prize setting if bidders'
preference relations satisfy the betweenness property[8]; ii) if the preference
relations on risky prospects are represented by real-valued functionals that
are non-linear in probabilities, then the equilibrium bidding strategy in
SPSB auctions is neither dominant nor value revealing.

The analysis of bidders' behaviour in SPSB auctions for random prizes
is conducted in Neilson (1994) who considers the case of implicit expected
utility maximising (IEUM) agents[9]. IEUM bidders have a continuously
Frechet differentiable preference function[10] $V(\cdot)$ which is defined as the
implicit solution to:

$$V(H) = \int u(z, V(H)) \, dH(z)$$

with $u(z; V(H))$ defined as the (single) local utility function, which is a
function of the payoff z and the final wealth distribution $H(\cdot)$.

Neilson (1994) shows[11] that there exists[12] a symmetric equilibrium
$B^*(\cdot)$, in which the optimal bid B_i^* is equal to the *conditional certainty
equivalent* of the random prize $X(r)$. The latter is the certainty equivalent
of the lottery $X(r)$ conditional upon the final wealth distribution $H(\cdot, \cdot)$.
This is the distribution of the random prize in the symmetric equilibrium
generated by $B^*(\cdot)$, when the bidder draws r_i from $G(\cdot)$ and bids B_i^*. The
conditional certainty equivalent solves the following:

$$\int_{X_a}^{X_b} u(X - B^*(r); V(H(\cdot, r, n, B^*, B^*(r)))) \, dF(X, r_i) = 0 \qquad (2)$$

---

[8] The latter is equivalent to the condition that the preferences over probability distributions be both quasi-concave and quasi-convex in the probabilities. Quasi-concavity
alone is necessary for the existence of a Nash equilibrium but it does not guarantee
revelation, since under (strict) quasi-concavity of preferences the bid is (strictly) higher
than the valuation, see Crawford (1990).

[9] In particular, bidders have smooth Implicit Expected Utility (IEU) preferences as in
Dekel (1986) and Chew (1989). The class of IEU preferences is narrower than the general
smooth preferences class considered in Machina (1982).

[10] Let $D[\alpha, \beta]$ indicate the set of all distribution functions over $[\alpha, \beta]$. Individuals are
assumed to have a complete and transitive preference relation $V(H)$, with $H \in D[\alpha, \beta]$.
Machina (1982) shows that, if the weak convergence topology is imposed on $D[\alpha, \beta]$,
$V(H)$ is Frechet differentiable.

[11] The intersted reader can look at Neilson's paper for the technical details while here
we only describe the intuition behind the results.

[12] The proof is based upon the assumption that the preference function $V(\cdot)$ is quasi-
concave in B_i.

From (2) we observe that any variable affecting the final wealth distribution $H(\cdot)$ influences the optimal bidding behaviour[13] $B^*(\cdot)$. In particular, each individual's behaviour is governed by changes in the shape of the local utility $u(\cdot;\cdot)$, which in turn is sensitive to the preference level $V(H(\cdot))$. Fanning hypotheses suggest how the local utility $u(\cdot;\cdot)$ is affected by changes of $V(H(\cdot))$. According to Machina (1982: H.II) fanning out hypothesis, if something happens that makes the agent better off, then his (Arrow-Pratt) degree of risk aversion increases. Therefore, under fanning out, an higher preference level implies a decreasing conditional certainty equivalent. The reverse occurs under fanning in, as the individual becomes less risk averse at increasing preference levels.

The revelation property of English auctions is maintained with IEUM bidders, since IEU theory satisfies the betwenness property. Therefore bidders have a dominant strategy in which they bid up to the certainty equivalent of the random prize, as in the EU case.

Using this result, Neilson (1994: Proposition 6) shows that IEU bidders satisfying fanning in (out) will submit an higher (lower) bid in the SPSB than in the English auction. As a consequence, the presence of bidders satisfying fanning in would generate a higher revenue in the SPSB auction with respect to the English auction.

The intuition behind this theoretical result is the following. An individual i is better off in a second price auction in which he bids B_i because he obtains a positive profit if winning. If he bids his certainty equivalent B_i in the English auction and wins, he has zero profit. Hence, the distribution of the final payoff in the SPSB auction dominates that of the English auction. This explains why an IEU bidder satisfying fanning in is made less risk averse when bidding in the SPSB auction with respect to the English auction.

We can then state the first set of experimental hypotheses as follows:

H.1 $B_E = B_{SPSB} \iff$ Expected utility maximiser

H.2 $B_E < B_{SPSB} \iff$ Implicit Expected utility maximiser with fanning in

H.3 $B_E > B_{SPSB} \iff$ Implicit Expected utility maximiser with fanning out

The above theoretical predictions have been investigated in our experiment contrasting the actual bids and the prices formed in a series of paired English and SPSB auctions conducted with the same set of private

---

[13] Since $H(\cdot)$ depends upon the other bidders' strategies, the bidding functions form a Nash equilibrium. Moreover the optimal bid is sensitive to the number of bidders.

random lotteries. This provide a test for the individual bidders behaviour in auctions. If bidders do not behave according to H.1, we should expect a failure of the revenue equivalence. However, this may not be necessarily true. The failure of EU theory in explaining individuals' behaviour, does not exclude a success in explaining the aggregate (or market) behaviour. The performance of EU theory at market level can be evaluated using the price series obtained from the two procedures. Our second set of experimental hypotheses is the following:

P.1 : $P_{SPSB} = P_E \iff$ Revenue Equivalence

P.2 : $P_{SPSB} \lesseqgtr P_E \iff$ Failure of Revenue Equivalence.

Our findings can be related to the results obtained in the experimental analysis of private value auctions with deterministic payoffs. A recurrent finding in this literature is that bidding behaviour in the English procedure quickly converges towards equilibrium[14]. On the contrary, agents appear to bid above the dominant strategy equilibrium in laboratory SPSB auctions[15]. Theoretical explanations of the latter experimental finding are hard to find: when the auctioned object has a deterministic value to bidders, bidding above the dominant strategy in SPSB can be only labeled as a "mistake" with respect to the (dominant) equilibrium play. Kagel (1995) conjectures that overbidding may be based on the bidders' illusion that it improves their probability of winning with little expected cost. The experimental results seem to confirm this hypothesis since the observed overbidding appears to be sustainable. This means that average profits remain positive for winners.

In our random prize environment two different explanations have to be considered. In the first place, the overbidding in the SPSB auction may be due to a fanning in-type of bidders' preferences. In the second place, overbidding can be motivated by an outside-the-equilibrium behaviour, for given bidders' preferences. To separate the two effects, we need some evidence on how bidders evaluate the random prizes outside the auction. This is done using a methodology proposed by Neilson (1992) which is described in the next subsection.

---

[14] This result has been obtained in both affiliated private value (Kagel et al., 1987) and in IPV models (Coppinger et al., 1980).

[15] Kagel and Levin (1993) prove the result both for experienced and inexperienced bidders having IPV valuations.

2.1. THE MODIFIED BECKER, DE GROOT, MARSCHAK DEVICE (MBDM)

For our comparative purposes we need to test bidders' attitudes toward risk when they have to attach monetary values to the random auction prizes in non-market conditions, i.e. when, ceteris paribus, we take aside the bidding competition. The information we need is hard to obtain when, as in the case of auctions, we are dealing with probability spaces involving more than three outcomes. Even if the random prize $X(\cdot)$ is represented by a two outcomes lottery, the final profit for the bidder in the SPSB procedure depends upon the second highest bid, which is a continuous random variable. As a consequence, the use of standard pairs of questions in the probability triangle does not seem to be appropriate to test for fanning behaviour.

Becker *et al.* (1964) proposed an incentive compatible mechanism to elicit the certainty equivalent of a lottery X. In brief, the agent is asked to specify an amount of money C and then he observes the realisation d of a random variable $\tilde{D}$, independent from X. If $d \geq C$ the agent receives d for sure, whereas if $C > d$ the agents obtains the random reward X. An expected utility maximiser would then choose C so that $u(C) = \int u(X) dF(X)$, and hence C is the certainty equivalent of the lottery X. The above result however holds if the agent is a true EU maximiser[16]. Assume on the contrary that he acts as an IEU maximiser instead. In this case, he would declare a value C^* that solves:

$$u\left(C^*; V\left(H\left(\cdot, C^*\right)\right)\right) = \int u\left(X; V\left(H\left(\cdot, C^*\right)\right)\right) dF(X)$$

so that now C^* represents the certainty equivalent of X *conditional* upon the final wealth distribution $H(\cdot)$. As a consequence, any change in the random variable $\tilde{D}$ affects $H(\cdot, C)$ and therefore $u(\cdot; H(C^*))$ changes.

The latter observation allowed Neilson (1992) to introduce a method for testing fanning properties in spaces more general than probability triangles. The test amounts at eliciting a value C_1 for a random lottery X, using an alternative random variable $\tilde{D}_1$, independent from X. Then the test is repeated with a second alternative random variable $\tilde{D}_2$ designed in a way that the agent is made better off at $H(C_1)$. If the second-elicited (conditional) certainty equivalent C_2 is less that C_1 then the individual has become more risk averse, a fact that suggests fanning out. On the contrary,

---

[16] More generally, the BDM procedure has been largely criticized by the literature. Karni and Safra (1987) concluded that the BDM mechanism elicits the certainty equivalent of a lottery only when the agent's preference relation can be represented by an expected utility functional.

if $C_2 > C_1$, then there is evidence of fanning in. Finally, $C_1 = C_2$ implies that the agent is (at least locally) an expected utility maximiser.

Bohm *et al.* (1997) applied the BDM mechanism in order to elicit selling prices for a commodity having a deterministic value to participants. The experimental findings show that the BDM mechanism is sensitive to the choice of the upper bound of the randomly generated buyout prices. In particular, ask prices are significantly higher when the upper bound for the bid price is increased. The above BDM experiment can be interpreted for our purposes by Neilson's point of view. Average asks increasing with the upper bound of the random bids indicate a tendency towards a higher certainty equivalent associated with a preferred final payoff distribution. This is in turn evidence of fanning-in.

In the light of this results, together with some encouraging evidence obtained from a pilot[17] we decided to apply the modified (or augmented) BDM methodology to obtain some outside-the-auctions evidence of fanning behaviour (if any)[18]. Details about the experimental design are contained in the next section.

3. The experimental design

The experiment involved a total of 106 participants recruited through the EXEC at the University of York (UK). Participants registered for one of a total of 15 experimental sessions, indexed by t, $t = 1, ..., 15$. Each session was divided into four parts and comprised a total of 18 auction rounds (9 couples of paired English and SPSB, indexed by $j = 1, .., 9$) spaced out by 20 pair-wise choice questions and 8 couples of modified-BDM questions. The experiment was completely computerised and the four parts of it were all linked together in a way that will be explained below[19].

In the two auction parts (Part 1 and 4 respectively) bidders competed for the right to play a "personal lottery" which represented the risky auction prize. Independent private values have been experimentally induced as follows. We started announcing a basic two outcomes lottery (equal for all the participant) in the form:

$$[\pounds Y, \ p; \pounds K, (1-p)]$$

---

[17] A pilot (non-computerised) experiment has been conducted at the University of Milan with non financially motivated participants. Pilot results are described in Parisio (1997).

[18] In this experiment, we do not use BDM mechanism as an incentive compatible device: we are only interested to the way the conditional certainty equivalent changes due to changes in $H(\cdot)$ and not to the actual values of C_1 and C_2.

[19] Instructions are available from the Author upon request.

TABLE I. Experimental Parameters

auction round (j)	1	2	3	4	5	6	7	8	9
personal bonus (r)	r_1	r_2	r_3	r_4	r_1	r_2	r_3	r_4	r_4
probability (p)	1/2	1/2	1/2	1/2	3/4	3/4	3/4	3/4	1

where Y and K represented a high and low monetary outcome respectively with the associated probabilities $p, (1 - p)$. Then each of the bidders privately observed a personal bonus r_{ij}, for bidder i in round j. Each bonus was obtained as a random draw from an interval of monetary values $[a, b]$ equal for all bidders. Hence for example, the following personal lottery PL_{ij} :

$$PL_{ij} = [\pounds (Y + r_{ij}), \; p; \pounds (K + r_{ij}), (1 - p)]$$
$$= [\pounds H_{ij}, p; \pounds L_{ij}, (1 - p)]$$

represents the auction prize privately known by bidder i in round j. During the 9 English auction rounds the experimental parameters varied as shown in TABLE I, whereas Y and K and the interval $[a, b]$ remained fixed through all the experimental session.

As we can see from TABLE I, in auction rounds 5 to 8 we repeated lottery outcomes of rounds 1 to 4, but we increased the probability of the best outcome from $\frac{1}{2}$ to $\frac{3}{4}$. Each couple of English auctions 1-5, 2-6, 3-7, 4-8, can be then interpreted as a way to elicit the certainty equivalents of two prospects which can be represented in the same probability triangle. The stop out price of bidder i in round j, P_{ij}, can be considered as the amount of money that the bidder is willing to pay such that the point $(0, 1, 0)$ and the point $((1 - p), 0, p)$ in the triangle lie on the same indifference line. At P_{ij} bidder i is indifferent between winning and losing the auction and hence it is the certainty equivalent of the prize elicited by the English auction mechanism. In English and SPSB auction round 9 the prize for bidder i was represented by a certain outcome $(p = 1)$ equal to $\pounds (Y + r_{i4}) = \pounds H_{i4}$. In this manner, we had the chance of testing whether bidders converged to the dominant strategy equilibrium of bidding an amount $\pounds H_{i4}$, both in the English and in the SPSB auction.

The nine SPSB auction rounds were performed in the fourth part of the experiment using for each bidder the same set of parameters shown in TABLE I. We choose to run the two auctions separately instead of asking bidders two different bids for the same prize, one associated to the English auction rules and the other to the SPSB auction. We think bidders can get (at best) confused and (at worst) influenced in their bidding decision when they have to announce their prices with alternating auction rules.

TABLE II.   Datasets

Dataset	K	Y	$[a, b]$	ε
DS1$(t = 1, ..., 5)$	£1	£5	£$[0, 4]$	£1
DS2$(t = 6, ..., 12)$	£0	£6	£$[1.5, 4.5]$	£1.5
DS3$(t = 13, ..., 15)$	£0	£8	£$[2, 4]$	£2

For our experiment we used three datasets for values of K, Y and of $[a, b]$. Details about datasets are shown in TABLE II.

The second part of the experiment was designed to evaluate the possibility of using an IEU preference functional to test participants' behaviour toward risk. We performed such test using 20 pair-wise choice questions which have been purposely generated by the computer program using the English auction data. More precisely, at each question a couple of lotteries taking the form of circles[20] appeared on the computer screen. Participants had to individually choose which of the two lotteries-circles he/she would prefer to play. The first group of five questions were related to English session 1, whereas questions 6 to 10 were related to English session 2 and so on.

Consider for example the first group of questions (1 to 5). Suppose that in English round 1 participant i dropped out at a price P_{i1}. In this way, he appears to be indifferent between an outcome of £0 with probability 1, and the following lottery:

$$[£ (H_{i1} - P_{i1}), p; £ (L_{i1} - P_{i1}) (1 - p)]$$

For each English round j, $(j = 1, ..., 4)$, the triple £ $(H_{ij} - P_{ij})$, £0 and £ $(L_{ij} - P_{ij})$ represents outcomes x_1, x_2, x_3 which can be associated to a probability triangle as the one represented in Figure 1.

Each couple of circles is designed to represent the choice between the following alternatives: 1) $A \succeq_{\prec} B$; 2) $A \succeq_{\prec} C$; 3) $A \succeq_{\prec} D$; 4) $A \succeq_{\prec} E$; 5) $E \succeq_{\prec} C$.

In this manner, Q1-Q2 and Q3-Q4 implement standard tests for betweenness, whereas Q5 is a test for stochastic dominance.

The other 15 pair-wise choice questions were constructed using bids P_{ij} and lotteries defined in English auction rounds $j = 2, 3, 4$.

In the third part of the experiment we performed 8 couples of modified BDM questions. For each participant we put on sale the personal lotteries of auction rounds 1 to 8, under a different exchange procedure: bidders did

---

[20] The circle part of the experiment has been designed following Hey and Orme (1994). Each circle represents a lottery with given monetary outcomes. Circles are divided into slices and the size of each slice represents the probability associated with each outcome.

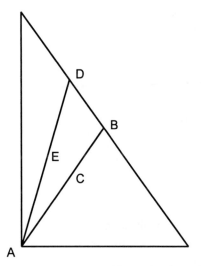

Figure 1. Lotteries in the circle questions

not compete in a market setting but they faced a random device instead. They were asked to state the maximum price D_1 at which they would be willing to buy (and play) the lottery: $[£H_{ij}, p; £L_{ij}(1-p)]$. The demand price D_1 had to be contrasted with an offer price $\tilde{N}$ randomly selected from the interval $[£L_{ij}, £H_{ij}]$. Then the following applies: $N_1 > D_1$: the bidder gets nothing $N_1 \leq D_1$: the bidder plays the lottery and pay the price N_1. Then each bidder was asked another demand price D_2 under a different offer perspective: $\tilde{N}$ is a random draw from the interval $[£(L_{ij} - \varepsilon), £H_{ij}]$, where ε is a positive value as illustrated in TABLE II. In this second question the bidder was made "better off" as he faced a higher probability of buying the same lottery at a lower price.

Participants were financially motivated. At the end of the experiment one part (1, 2, 3 or 4) was selected at random and then again we selected at random one round from that part. The selected round was then played for real and participants paid according with the actual outcomes. Participants received a fixed fee of £10 for a total of 50 minutes of task, plus (minus) the profits (losses) earned in the round played for real. The experimental parameters illustrated in TABLE II have been designed so as to avoid the possibility of bankruptcies.

4. The experimental results

We first classify participants on the basis of their individual bidding behaviour in the two auction procedures. To this end, in TABLE III, we

TABLE III.   Individual bidding behaviour in auctions

Group	Behaviour	Bidders
A	H.1	5.4%
B	H.2	38.5%
C	H.3	16.6%
D	H.1 rounds 1-4/H.2 rounds 5-8	11.2%
E	H.1 rounds 1-4/H.3 rounds 5-8	7.3%
F	H.2 rounds 5-8/H.3 rounds 5-8	11.9%
G	H.3 rounds 5-8/H.2 rounds 5-8	0
H	MIX of H.1, H.2, H.3	9.1%

allocate bidders to one of eight groups, respectively identified as Group A, B, C, D E, F, G and H in the first column. The second column of TABLE III indicates the typical bidding behaviour that characterises individuals belonging to the group, so that, for example, we allocate to group A, B and C respectively bidders who are consistent with our experimental hypotheses H.1, H.2 and H.3. In the same manner, we allocate to group C, D, E, F bidders who change their bidding behaviour from the first four auction rounds (where the lottery at stake had 1/2 chance of the two outcomes) to the second four auction rounds (where we increased the probability of the better outcome to 3/4). Finally, we allocate to group H bidders who followed a mixed type of bidding behaviour. In the third column of TABLE III, we indicate the number of bidders (% of the total) belonging to the group.

Looking at the auction results, we see that the highest percentage of agents (38% from TABLE III) seems to behave following the fanning-in hypothesis H.2, which implies an higher bid in the SPSB auction with respect to the English procedure. A very low proportion of bidders behave on the basis of the Expected Utility hypothesis H.1. Group D, showing a mix of EU and fanning in-type of behaviour account for around 11% of bidders. Fanning out or a combination of EU/fanning out is found for a total of around 25% of participants. Some agents (group G) appeared to follow a mixed type of behaviour oscillating between fanning-in, fanning-out or EU. They account for around 9% of the total. Finally, a proportion close to 12% of bidders appeared to follow a mixed behaviour in the sense that they showed to be consistent with fanning-in in auction rounds 1 to 4, whereas they switched to fanning-out when they bid for stochastically dominating lotteries (rounds 5 to 8). No bidder appeared to be consistent with the reverse pattern, i.e. from fanning-out to fanning-in (see Group G).

The low proportion of agents that has been found to behave according to the expected utility hypothesis seems to suggest that, at the market

TABLE IV.   Descriptive statistics of price series

	English	SPSB
Mean	7.45	7.04
Median	7.20	7.00
Max	10.30	11.00
Min	2.20	3.55
St. Dev	2.05	1.39
Sk.	0.26	0.37
Kurt.	2.91	3.04
n. obs.	135	135
t-test	P-val.	
0,004323	0,996557	

level, we should reject revenue equivalence (hypothesis P.1). Therefore, it is interesting at this stage to compare the price levels obtained in our experimental English and SPSB auctions. TABLE IV reports some descriptive statistics of the two series of prices. The average price obtained in the English procedure appears to be slightly larger than his SPSB auction counterpart. English prices are also more dispersed around the mean with respect to SPSB auction prices. In the last two rows of TABLE IV we report the result of a paired t-test conducted using our price series and based on the following null and alternative hypotheses:

$$H_0 = P.1 : \Delta P = P_E - P_{SP} = 0$$
$$H_1 = P.2 : \Delta P = P_E - P_{SP} \neq 0$$

The t-test conducted over the paired observations of all the sessions appears to accept the null hypothesis of $P_E = P_{SP}$. As a consequence, the experimental data are consistent with the theoretical prescription of our expected utility hypothesis P.1. This conclusion is quite important since the violations of expected utility observed at the individual bidding level appears to be smoothed down at the market level. The latter experimental finding is similar to that of Evans (1997) but our result appears to be stronger because it is obtained under a second-price rule which obviously induces a less severe bidding selection procedure with respect to the fifth price auction rule used by Evans.

The individual bidding behaviour and the effect of competition on auction prices can be further investigated by means of the result obtained in the modified BDM part of our experiment. TABLE V reports the allocation of participants into the same eight groups of TABLE III on the basis of their

TABLE V.   Individual behaviour in the MBDM

Group	Behaviour		MBDM
A	$D_1 = D_2 \Leftrightarrow EU$		20.7%
B	$D_1 > D_2 \Leftrightarrow$ fanning in		8.9%
C	$D_1 < D_2 \Leftrightarrow$ fanning out		28.3%
	$p = 1/2$	$p = 3/4$	
D	$D_1 = D_2 \Leftrightarrow EU$	$D_1 > D_2 \Leftrightarrow$ fanning in	15.7%
E	$D_1 = D_2 \Leftrightarrow EU$	$D_1 < D_2 \Leftrightarrow$ fanning out	17.4%
F	$D_1 > D_2 \Leftrightarrow$ fanning in	$D_1 < D_2 \Leftrightarrow$ fanning out	0.9%
G	$D_1 < D_2 \Leftrightarrow$ fanning out	$D_1 > D_2 \Leftrightarrow$ fanning in	0
H	mixed		8.1%

MBDM behaviour. In the second column, groups from D to G include those bidders who changed their behaviour when they evaluated the stochastically dominating lotteries, that is the same lotteries that they encountered in auction rounds 5 to 8.

Looking to the modified BDM results of TABLE V, we notice that the largest proportion of agents appeared to follow a fanning-out type of behaviour (group C). In the same manner, a good proportion of agents (17, 4%) appear to switch from EU to fanning-out when they had to attach values to the preferred lotteries (group E). On the contrary, the proportion of agents consistent with fanning-in (group B) or switching from EU to fanning-in (group D) is found to be fairly poor. Surprisingly, we notice that a very large number of participants followed an EU type of behaviour in the modified BDM part of the experiment. This means that the proportion of expected utility violations is strongly less when the bidding competition is eliminated.

A closer look to the experimental data confirms the latter finding. TABLE VI, written in matrix form, has the auction data arranged on the basis of the top (label) row, while MBDM data are arranged in column. So, for example, the first column indicates that of the 6 participants consistent with expected utility theory in auctions, 4 behave as EUM even in the MBDM part whereas 1 was consistent with fanning-in and 1 was consistent with fanning out.

Looking at the second column, it is evident that a good number of participants switched from fanning-in at auctions toward and EU type of behaviour in the MBDM part[21]. Moreover, a number of 13 bidders turned

---

[21] This result is opposite to the one obtained by Evans (1997) a fact that can be explained on the basis of the different pricing rules adopted.

TABLE VI.   Auction and MBDM behaviour

		AUCTION								
		A	B	C	D	E	F	G	H	TOT.
M	A	4	8	3	2	3	1	0	1	22
B	B	1	8	0	0	0	1	0	0	10
D	C	0	13	10	2	1	5	0	0	31
M	D	0	6	2	5	0	3	0	0	16
	E	1	5	2	1	4	2	0	3	18
	F	0	0	0	0	0	1	0	0	1
	G	0	0	0	0	0	0	0	0	0
	H	0	1	1	1	0	0	0	5	8
	TOT.	6	41	18	11	8	13	0	9	106

from fanning-in at auctions to fanning-out at the modified BDM questions.

The strong occurrence of fanning-in shown by participants only when they compete at auctions, might be explained by the effect on bidders' strategies of the second-price rule. As Kagel (1995) pointed out for the deterministic case, agents may think that overbidding in a second price auction with respect to the English auction, increases their probability of winning at a negligible cost. We can first investigate the validity of this intuition looking at the results obtained in the certainty rounds (round 9) of both English and SPSB auctions. In these rounds, we found a good proportion of bids consistent with the dominant strategy, which requires $P_E = P_{SP}$, but, on the other hand, we also found evidence of some outside-the-equilibrium behaviour. In particular, we noticed underbidding in the English procedure with respect to the SPSB. The most interesting results however, is that a good proportion of the agents who appeared to follow a fanning-in type of behaviour in auction rounds 1 to 8 (with risky prize), overbid in the certainty round (round 9) of the second price auction. This finding, together with the evidence that auction competition increases the evidence for fanning-in with respect to what happens in an incentive compatible mechanism as the MBDM, confirms us that there should be some other strategic factor that influences agents' behaviour in the second price auction. We therefore conclude that preferences alone cannot explain the bidding behaviour at the SPSB auction with respect to the English auction. It might be that participants attach some value to the fact of winning, or as put forward by Kagel, they expect to increase their probability of winning at a little cost. Our findings seem to confirm the latter hypothesis, since the overbidding at SPSB auction was limited to reasonable terms and in no cases it lead to negative profits. Moreover, as already illustrated in TABLE IV, the effect of overbidding disappeared at the price level.

In the second part of the experiment we tested by means of circle questions whether or not bidders' preferences over lotteries satisfy the requirements of betweenness and of stochastic dominance. Experimental results[22] indicate that violations of the two assumptions were fairly negligible. The average rate of violation across groups was around 3.8% for betweenness and 1.5% for stochastic dominance. Interestingly, we notice that the strongest rate of violations occurred for bidders showing a mixed type of behaviour (Group H) or for bidders who alternate fanning-in and fanning-out (Group F)[23]. Given these results, we can (ex-post) justify the use of an IEU preference functional for our experimental predictions.

5. Conclusions

Recent developments in the theory of auctions with random payoffs and without Expected Utility provide a series of behavioural prescriptions which can be investigated by means of experimental methods. In this paper we presented the results of an experiment that took place at the EXEC, (University of York, UK) and involved a number of 106 participants. In each of the 15 experimental sessions we conducted a paired series of English and Second price auctions for random lotteries. We found that the price series generated by the two mechanisms are likely to come from the same two populations having equal mean. The latter result is the only one consistent with the revenue equivalence prescribed by expected utility theory. Looking at the individual bidding level we found on the contrary a not negligible rate of expected utility violations in a direction typical of fanning-in preferences. Such violations however, appear to be smoothed by the market selection rule which, in both auctions picks the second-highest bid price.

In the third part of our experiment we further investigated the effect of bidding competition on agents behaviour. We conducted a test of bidders' behaviour towards risk using an augmented version of the Becker, De Groot and Marshack (BMD) mechanism, first proposed by Neilson (1992). Coupling auction data with the modified BDM results we found that bidders' behaviour is sensitive to the environment which is used to solicit their certainty equivalent of their random lotteries. We was able to recognise a general pattern of behaviour which emphasises the occurrence of fanning-in at auctions and of fanning out or an Expected Utility type of behaviour in the modified BDM mechanism. All together it seems that market competition induces a higher rate of EU violations at the individual bidding level with respect to an incentive compatible treatment were bidding competition is eliminated.

---

[22] Full experimental data are available from the Author upon request.

[23] In particular, we notice that the rate of violation of betweenness was around 18% in Group H and 25% in Group F.

References

Becker G. M., M. H. DeGroot and Marschak J.: 1964, 'Measuring Utility by a Single-Response Sequential Method', *Behavioral Science*, **9**, pp. 61-104.

Bohm P., Lindén J., and Sonnegård J.: 1997, 'Eliciting Reservation Prices: Becker-Degroot-Marschak Mechanisms vs. Markets', *The Economic Journal*, **107**, pp. 1079-1089.

Chew S.H.: 1989, 'Axiomatic Utility Theories with the Betweenness Property', *Annals of Operation Research*, **19**, pp. 273-298.

Coppinger V. M., Smith V. L. and Titus J. A.: 1980, 'Incentives and Behavior in English, Dutch and Sealed-bid Auctions', *Economic Inquiry*, **43**, pp. 1-22.

Crawford V.: 1990, 'Equilibrium without Independence', *Journal of Economic Theory*, **50**, pp. 127-154.

Dekel E.: 1986, 'An Axiomatic Characterisation of Preferences under Uncertainty: Weakening the Independence Axiom', *Journal of Economic Theory*, **40**, pp. 304-318.

Evans D. A.: 1997, 'The Role of Markets in Reducing Expected Utility Violations', *Journal of Political Economy*, **105(3)**, pp. 622-636.

Hey J. D. and Orme C.. 1994, 'Investigating Generalizations of Expected Utility Theory using Experimental Data', *Econometrica*, **62(6)**, pp. 1291-1326.

Kagel, J. H.: 1995, 'Auctions: A Survey of Experimental Research', in Kagel, J. H. and Roth A. E.(eds.), *The Handbook of Experimental Economics*, Princeton University Press, Princeton (U.S.).

_____, R. M. Harstad and Levin D.: 1987, 'Information Impact and Allocation Rules in Auctions with Affiliated Private Values: A Laboratory Study', *Econometrica*, **55**, pp. 1275-1304.

_____ and Levin D.: 1993, 'Independent Private Value Auctions:Bidder Behavior in First-, Second- and Third-price Auctions with Varying Number of Bidders', *Economic Journal*, **103**, pp. 868-879.

Karni, E. and Safra Z.: 1986, 'Vickrey Auctions in the Theory of Expected utility with Rank Dependent Probabilities', *Economic Letters*, **20**, pp. 15-18.

_____, _____: 1987, 'Preference Reversal and the Observability of Preferences by Experimental Methods', *Econometrica*, **55(3)**, pp. 675-685.

_____, _____: 1989, 'Dynamic Consistency, revelations in Auctions and the Structure of Preferences', *Review of Economic Studies*, **56**, pp. 421-434.

Machina M.: 1982, 'Expected Utility Analysis without the Independence Axiom', *Econometrica*, **50**, pp. 277-323.

McAfee R. P. and McMillan J.: 1987, 'Auctions and Bidding', *Journal of Economic Literature*, **25**, pp. 699-738.

Milgrom P. R. and Weber R. J.: 1982, 'A Theory of Auctions and Competitive Bidding', *Econometrica*, **50(5)**, pp. 1089-1122.

Neilson W.: 1992, 'A mixed Fan Hypothesis and its Implications for Behavior Towards Risk', *Journal of Economic Behavior and Organization* **19**, pp. 197-211.

_____: 1994, 'Second Price Auctions without Expected Utility', *Journal of Economic Theory*, **62**,pp. 136-151.

Parisio L.: 1997, 'A pilot Experiment on Auctions with Random Prizes and EU/non-EU Bidders', Annual Meeting of Experimental Economics, University of Trento, 6-7 June 1997.

Wolfstetter E.: 1996, 'Auctions: an Introduction', *Journal of Economic Surveys*, **10(4)**, pp. 367-420.

CHAPTER 13

DYNAMIC GAMES AND OLIGOPOLY MODELS OF TECHNOLOGICAL INNOVATION

MARIA LUISA PETIT (petit@dis.uniroma1.it)
Dipartimento di Informatica e Sistemistica
University of Rome "La Sapienza"
via Buonarroti 12
00185 Rome, Italy

BOLESLAW TOLWINSKI
ORE (Operations Research Experts)
Golden, Colorado 80401, USA

Abstract. Since innovation is clearly a dynamic phenomenon, the process of technological innovation should be analysed by making use of a dynamic approach. In this paper we present two different models of innovation in the framework of an oligopolistic market and show that differential (or difference) games can provide an appropriate analytical tool to analyse this kind of problems. In this context it is possible to introduce state variables representing the technological knowledge accumulated by the firms over time and to link the innovation process to those variables. Equilibria in Markov strategies are computed by using a modified policy iteration algorithm.

1. Introduction

The process of technological innovation and its diffusion within the industrial system are *dynamic* processes and should therefore be analysed by making use of a dynamic approach. In this paper we wish to show that a particular class of dynamic games called differential (difference) games or Markov games are an appropriate tool to analyse the process of innovation, since they make it possible to consider not only a decision making problem structured over a given time horizon, but also an *economic environment that evolves over time.* We believe in fact that innovation is the result of knowledge, and in particular of *accumulated* knowledge, since the past performance of the firm (and of the industry) is a fundamental element of current behaviour. Process and product innovations produced by a firm are directly related to the level of knowledge that it has been able to reach over time. Or, in the case of an economy with technological spillovers, also to the level of knowledge that the other firms producing in the same or in close industrial sectors have been accumulating in the past. Differential games make it possible to introduce a state variable whose behaviour over time can represent the evolution of accumulated knowledge.

The need for a dynamic approach to analyse the process of innovation was clear since the pioneering work of Kamien and Schwartz (1982). The more recent literature has followed this approach by making use of supergames, evolutionary games and also differential games (though in a completely different framework form the one presented here. See Reinganum,1990, for a survey).

The approach followed in this paper makes also possible to examine the evolution of possible asymmetries between the firms, and thus the evolution of the industrial structure. Furthermore, the equilibrium solutions that we shall consider (i.e. Markov perfect equilibria), obtained by applying dynamic programming techniques, have the property of being subgame perfect in the sense of Selten (1975) (or strong time consistent, as defined by Basar and Olsder, 1995).

We shall assume that knowledge is "produced" in a continuous form. This means considering innovations not as radical innovations that take place at given isolated and uncertain moments of time, but as a flow of incremental innovations which take the form of a sequence of improvements in products and production processes. Although both forms of innovations obviously coexist, a world of rapid evolving technology is more often characterized by innovations of the latter type.

Since the main source of innovation is *learning* by firms, we shall consider two forms of learning: learning by doing and learning from R&D.

Learning by doing can be represented by the learning (or experience) curve introduced by Arrow (1962). The idea behind the experience curve is well known: unit production costs of a firm decrease as accumulated production increases over time. For this reason, production can be considered as a means for increasing experience; that is, the acquisition of experience can no longer be seen as an unintentional byproduct of production but also as the result of an *investment decision* in learning. Accumulated knowledge is thus described by the levels of the accumulated production of the firms, which are the state variables of the model.

The basic assumption of *learning from R&D* is that R&D investments give rise to a flow of knowledge that transforms itself into a flow of process innovations. This has been formalized in the literature by assuming that unit production costs are negatively related to current R&D investment (see, e.g. D'Aspremont and Jacquemin,1988). The use of a dynamic model makes it possible to relax the very restrictive assumption that production costs decrease with *current* expenditure in R&D, and to assume that they decrease with *accumulated* investment in R&D. Accumulated knowledge is thus described by the levels of R&D capital accumulated by the firms over time, which are the state variables of the model.

The paper is organized as follows. In section 2 a brief survey of difference games and Markov perfect Nash equilibria are presented. Section 3 describes the different duopoly models that can be used to represent the different forms of learning by the firms. In this section, some specific forms of technological agreements are also introduced. In section 4 the concept of Markov equilibrium is applied to the different models. Numerical examples concerning the case of asymmetric firms are presented in section 6. Section 7 provides some conclusive remarks. An Appendix describes the numerical algorithm.

2. Dynamic Games and Markov Equilibria

We recall here the main elements that characterize a difference game, the conditions for Markov perfect Nash equilibria and the technique (dynamic programming) for deriving such equilibria. Since the models that follow will be formulated as discrete-time games, only this class of games will be considered. (For a wider analysis see Basar and Olsder, 1995).

2.1. ELEMENTS OF A DYNAMIC (DISCRETE-TIME) GAME

An N-person discrete-time deterministic game involves
   (i) an index set $\mathbf{N}$ (the player set), i.e. $\mathbf{N} = \{1, 2, ..., N\}$.
   (ii) an index set $\mathbf{T} = \{0, 1, ..., T-1\}$ denoting the stages of the game, where T is the maximum possible number of moves a player is allowed to make in the game. $\mathbf{T} = \{0, 1, ..., \infty\}$ is an infinite horizon game.
   (iii) an infinite set X called the *state set (space)* of the game, to which the state of the game (x_t) belongs for all $t \in \mathbf{T}$.
   (iv) an infinite set U_t^i defined for each $t \in \mathbf{T}$ and $i \in \mathbf{N}$, called the *action set* of player i at stage t.
   (v) a function $f : X * U_t^1 *, \dots , *U_t^N * \mathbf{T} \to X$, so that

$$x_{t+1} = f(x_t, u_t^1, \dots , u_t^N, t) \qquad (1)$$

for some vector $x_0 \in X$ which is called the initial state of the game. The system of difference equations (1) is called the *state equation* of the dynamic game and describes the evolution of a dynamic system over time.

(vi) a finite set η_t^i defined for each $t \in \mathbf{T}$ and $i \in \mathbf{N}$ as a subcollection of $\{x_0, x_1, \ldots, x_t; u_0^1, u_1^1, \ldots, u_{t-1}^1; u_0^2, u_1^2, \ldots, u_{t-1}^2; \ldots ; u_0^N, u_1^N, \ldots, u_{t-1}^N\}$ which describes the information of player i at stage t of the game. Specification of η_t^i for all $t \in \mathbf{T}$ characterizes the *information structure* of player i (i.e. $\eta^i = \{\eta_t^i : t = 0, 1, \ldots, T-1\}$). The *information structure* or *information pattern* of the game is defined by the collection of the information structures of all the players, that is $\eta = \{\eta^1, \ldots, \eta^N\}$.

(vii) a set Γ^i of permissible strategies γ^i of player i, defined for each $i \in \mathbf{N}$. Γ^i is known as the strategy space of player i. At each stage t, γ_t^i associates an action u_t^i with every η_t^i (the information available at time t).

Let τ be any stage of the game, $\tau \geq 0$, then

$$u_\tau^i = \gamma_\tau^i(\eta_\tau) \quad \text{for all } i \in \mathbf{N} \tag{2}$$

(viii) an objective function J^i defined for each player $i \in \mathbf{N}$. J^i is called the payoff function of player i and can be defined in a variety of ways. An example is

$$J^i(\tau, \eta_\tau; u^1, \ldots, u^N) = \sum_{t=\tau}^{\infty} \rho^{t-\tau} g^i(x_t, u_t^1, \ldots, u_t^N) \tag{3}$$

for the infinite-horizon case, where ρ is a discount factor.

2.2. INFORMATION PATTERNS OF A DYNAMIC GAME

The solution of a dynamic game will be different depending on the assumptions made about the information structure of the players. The most common types of information patterns are

- *Open-loop information pattern.* At each stage t the players know only the initial state of the system, i.e.,

$$\eta_t = \{x_0\} \quad \text{for all } t \tag{4}$$

- *Feedback information pattern.* At each time t the players observe the current state of the system, i.e.,

$$\eta_t = \{x_t\} \quad \text{for all } t \tag{5}$$

- *Closed-loop with unobservable actions information pattern.* All players have access to the information about the current state of the system and recall its past values, i.e.,

$$\eta_t = \{x_0, x_1, \ldots, x_t\} \quad \text{for all } t \tag{6}$$

— *Closed-loop with observable actions information pattern.* Each player observes the state and the other player's actions. He has perfect recall of his past observations and actions, i.e.

$$\eta_0 = \{x_0\};$$

$$\eta_t = \{u_0^1, \ldots, u_0^N, x_0, \ldots, u_{t-1}^1, \ldots, u_{t-1}^N, x_{t-1}, x_t\} \tag{7}$$

for $t = 1, 2, 3 \ldots$

In each of the information patterns now defined it is assumed that the information vector η_t is common knowledge for all the players.

2.3. NASH EQUILIBRIA

Specification of the information pattern of a dynamic game makes it possible to define the corresponding strategy spaces and then, to define payoffs as functions of strategies (rather than actions). Thus, for any initial data $(\tau, \gamma(\tau))$, the dynamic game can be represented as a game in *normal form*, which makes the concept of Nash equilibrium readily applicable.

DEFINITION 2.1. An N-tuple of strategies $\gamma^* = (\gamma^{*1}, \ldots, \gamma^{*N})$ is a *Nash equilibrium* at a given point (τ, η_τ) if

$$J^1(\tau, \eta_\tau; \gamma^*) \geq J^1(\tau, \eta_\tau; \gamma^1, \gamma^{2*}, \ldots, \gamma^{*N})$$
$$J^2(\tau, \eta_\tau; \gamma^*) \geq J^2(\tau, \eta_\tau; \gamma^{1*}, \gamma^2, \ldots, \gamma^{*N})$$
$$\ddots$$
$$J^N(\tau, \eta_\tau; \gamma^*) \geq J^N(\tau, \eta_\tau; \gamma^{1*}, \gamma^{2*}, \ldots, \gamma^N) \tag{8}$$

The possibility that Nash equilibrium of a dynamic game may be independent from *initial data* leads to another, much stronger equilibrium concept.

DEFINITION 2.2. An N-tuple of strategies $\gamma^* = (\gamma^{*1}, \ldots, \gamma^{*N})$ is a *(subgame) perfect Nash equilibrium* of a dynamic game if it is an equilibrium *at every possible point* (τ, η_τ).

2.4. DYNAMIC PROGRAMMING

One of the most important mathematical tools for the computation of (sub-game) perfect Nash equilibria is dynamic programming. This technique, introduced in the context of single agent decision problems (i.e. optimal control problems), has been extended to the framework of dynamic game theory.

Consider an infinite horizon dynamic game with the players maximizing payoff functions

$$J^i(\tau, \eta_\tau; u^1, \dots, u^N) = \sum_{t=\tau}^{\infty} \rho^{t-\tau} g^i(x_t, u_t^1, \dots, u_t^N) \qquad (9)$$

subject to the system dynamics

$$x_{t+1} = f(x_t, u_t^1, \dots, u_t^N) \qquad (10)$$

To solve this problem we need to determine the so-called *Bellman functions* (also referred to as *value functions*, $W^i(x)$, that satisfy the following equations

$$W^1(x) = \max_{u^1} H^1(x; u^1, \gamma^2(x), \dots, \gamma^N(x))$$
$$W^2(x) = \max_{u^2} H^2(x; \gamma^1(x), u^2, \dots, \gamma^N(x))$$

$$\ddots$$

$$W^N(x) = \max_{u^N} H^N(x; \gamma^1(x), \gamma^2(x), \dots, u^N) \qquad (11)$$

where, for all $i \in \mathbf{N}$,

$$H^i(x; u_1, \dots, u_N) \equiv g^i(x, u^1, \dots, u^N) + \rho W^i(f(x, u^1, \dots, u^N)) \qquad (12)$$

and where $\gamma = (\gamma^1, \dots, \gamma^N)$.

Then the following theorem holds:

THEOREM 2.1. *Suppose that there exist value functions $W^i(x)$ and strategies $\gamma^i(x)$, $i \in \mathbf{N}$, that satisfy Bellman's equations as defined above. Then the strategy N-tuple $\gamma = (\gamma^1, \dots, \gamma^N)$ is a perfect equilibrium of the game. Moreover, the value function $W^i(x)$ represents the payoff of player i generated by γ, for any value of τ, when the game starts at (τ, x_τ).*

If γ_t^i does not explicitly depend on t, that is, all γ_t^i are identical for $t = 0, 1, 2, \ldots$, then the resulting strategy γ^i is called a *stationary strategy*. In such case, the symbol γ^i is often used to denote both player i's decision rule at period t and the whole infinite sequence of those rules. The importance of stationary strategies stems from the fact that games played over an infinite horizon often admit equilibria which can be expressed in terms of such strategies and, in view of Theorem 2.1, the equilibria of that type can be computed, at least in principle, as solutions to Bellman's equations.

The equilibria obtained via dynamic programming with the value functions defined as functions of the state vector x, are usually referred to as *feedback Nash equilibria* or *Markov equilibria*. A feedback equilibrium is a *perfect equilibrium* under any of the information patterns described above, even though the players use only information concerning the current values of the state vector.

The dynamic programming approach requires the determination of the value functions $W^i(x)$, for every $x \in X$. This can be achieved in practice (and not always) only if W^i is affine or quadratic in x. The latter occurs when the dynamic game has linear dynamics and quadratic payoff functions. In other, more complex, non-linear cases the problem is generally not solvable analytically, and even numerical solutions are hard to compute.

3.   Dynamic-game models of innovation and learning

The two forms of learning described in Sect. 1 are now formalized by using two different groups of models, even if these different models have many common characteristics. They are defined as infinite horizon discrete-time games, with non-linear demand and non-linear learning functions. It is assumed that there are two firms that produce a homogeneous good and operate with constant returns to scale technology at each period of time. The firms can be asymmetric, i.e. technology can be different for each firm and unit costs of production at any given time may differ from firm to firm. In both cases technological innovations are process innovations which result in reductions in unit costs of production. It is also assumed that innovations of the two firms are complementary, so that both firms will benefit from technological spillovers or from deliberately sharing information with each other. The specific characteristics of the two models are described in the next two subsections.

3.1.   LEARNING BY DOING

When learning by doing characterizes the production process the assumption is made that unit cost decreases as cumulative production of a firm increases. The level of production is the decision variable of each firm and each firm knows that increasing production increases experience, thus reducing unit costs. Learning is assumed to be firm specific and the learning process may differ from firm to firm.

Firms are indexed by i, $i = 1, 2$. We define $q_{it} \geq 0$ to be firm i's output at time t, and $w_{it} \geq 0$ to be its cumulative output (i.e., experience) from time 0 up to time t. Let $Q_t = q_{1t} + q_{2t}$ denote aggregate production at period t. The corresponding demand is defined by a nonlinear stationary inverse demand function of the constant elasticity type. More specifically

$$P(Q) = AQ^{-\beta} \tag{13}$$

where $Q = q_1 + q_2$, $\beta = 1/B$, and B represents the elasticity of demand.

The state variable w_{it} is related to the control variable q_{it} through the state equation

$$w_{it+1} = w_{it} + q_{it} \quad (i = 1, 2) \tag{14}$$

Let $c_i(w_i)$ denote firm i's constant marginal (unit) cost per period corresponding to the level of experience w_i; $c_i(w_i)$ also represents the *learning* or *experience curve*, that is the relation between unit cost and experience. As indicated before, the unit cost of a firm will be assumed to decrease as its cumulative output grows. More specifically, we shall consider learning curves of the form

$$c_i(\underline{w}) = c_i^o(1 + w_i + \alpha w_j)^{-D_i} + c_i^{min}, \quad i, j = 1, 2; \quad i \neq j, \quad 0 \leq \alpha \leq 1 \tag{15}$$

where $\underline{w} = (w_1, w_2)$ and where c_i^{min} is the minimum unit cost that firm i can achieve ($c_i(w_i)$ tends asymptotically to c_i^{min} as w_i tends to infinity). The initial unit cost of firm i, $c_i(0)$, is given by $c_i^o + c_i^{min}$, where c_i^o is the initial cost disadvantage of firm i relative to its long run minimum c_i^{min}. The parameter D_i determines the rate at which unit costs decline with accumulated production from $c_i^o + c_i^{min}$ toward c_i^{min}, or, in other words, the speed of learning. Notice that the firms are allowed to have different learning curves. The asymmetry can result from different initial and minimum unit costs, from different learning speeds, or from both.

The parameter $\alpha \in [0, 1]$ is a spillover parameter. This means that, if $\alpha > 0$, the magnitude of firm i's cost reduction at any time t is determined by its own accumulated technological knowledge up to t and by a fraction α of the knowledge accumulated by the other firm.

It is assumed that each firm maximizes the present value of profits over an infinite horizon [1], that is

$$\pi_i^k = \sum_{t=0}^{\infty} \rho^t l_i^k(w_{it}, q_{1t}, q_{2t}) \qquad (16)$$

where

$$l_i^k(w_i, q_1, q_2) = q_i \left[A(q_1 + q_2)^{-\beta} - c_i^o(1 + w_i + \alpha w_j)^{-D_i} - c_i^{min} \right] \qquad (17)$$

where ρ denotes a discount factor, $0 < \rho < 1$, and

$$l_i^k = \begin{cases} l_i^\sigma, & if \ 0 \le \alpha < 1 \\ l_i^{TSC}, & if \ \alpha = 1 \end{cases} \qquad (18)$$

The superscript σ is employed to distinguish the case of no spillovers or involuntary spillovers (π_i^σ) from the case in which information is *shared* (π_i^{TSC}).

The case of no spillovers ($\alpha = 0$) may arise when there is strong intellectual property protection, either through patents or because firms are able to keep their information secret. Frequently, however, involuntary information leaks occur as a result of reverse engineering, industrial espionage or by hiring away employees of an innovative firm (see, e.g. Mansfield, 1985). The case of $\alpha = 1$ describes the existence of a new phenomenon: the *voluntary* transmission of technological knowledge among oligopolistic firms competing in the same market (Katz and Ordover,1990; Baumol, 1993). The agreement to share technological knowledge does not mean that the firms disclose technical details that would have otherwise remained secret, but simply that they accelerate a process of diffusion of knowledge that would have taken place in any case[2]. Following the terminology used by Baumol (1993) we will refer to this form of agreement between firms as "Technology Sharing Cartels", or "Technology Sharing Consortia", or simply TSC.

Under this assumption, the learning curve can be described as

---

[1] The assumption of a game of infinite duration can be interpreted as either an approximation to a game of a long but undefined duration, or as a finite horizon game with random stopping time. See, e.g. Friedman (1991).

[2] Empirical research shows that patents are not generally an effective method of appropriability and rival firms normally learn about technical characteristics of new products and processes within 12 months of their introduction. See Mansfield (1985), Levin et al. (1987).

$$c_i(w) = c_i^o(1+w)^{-D_i} + c_i^{min}, \quad i = 1,2 \tag{19}$$

where the joint accumulated experience evolves according to

$$w_{t+1} = w_t + q_{1t} + q_{2t} \tag{20}$$

3.2. LEARNING FROM R&D

In this case it is assumed that the cost reducing technological innovations
are an outcome of the firms' accumulated capital in R&D. Thus current
R&D investment becomes a strategic element of the firms behaviour. Learning resulting from cumulative investment in R&D characterizes the production process, implying that unit costs decrease as the *cumulative* R&D
investment increases, approaching a minimum attainable value depending
on the amount of R&D accumulation of each firm over time.

Firms can be asymmetric, i.e. they may use different technologies and
their unit costs of production at any given time may differ. As before, the
firms are indexed by i, $i = 1,2$. We define $q_{it} \geq 0$ to be firm i's output at
time t, while $Q_t = q_{1t} + q_{2t}$ denotes the aggregate output at period t. The
corresponding demand is again defined by Eq. (13). Let $w_{it} \geq 0$ be firm i's
cumulative technological knowledge resulting from (and represented by)
R&D capital accumulation, i.e., the R&D capital accumulated from time 0
up to time t. Firm i's current investment in R&D is denoted by u_{it}.

The firms decide how much to produce and how much to invest in R&D.
Thus the two decision variables of firm i are q_{it} and u_{it}, $i = 1,2$. The state
variable w_{it} is, therefore, related to the control variable u_{it} through the
state equation

$$w_{it+1} = (1 - \mu)w_{it} + u_{it} \quad (i = 1,2) \tag{21}$$

where $\mu \in [0,1)$ is the rate of depreciation of R&D capital.

Let $c_i(w_i)$ denote firm i's marginal (unit) cost per period corresponding
to the level of knowledge w_i. The function $c_i(w_i)$ represents the relation
between firm i's unit cost and its level of technical knowledge. As indicated
before, the unit cost of a firm will be assumed to decrease as its cumulative
R&D investment grows. Since we also allow for the possibility of technological spillovers between firms, we introduce a spillover parameter $\alpha \in [0,1]$,
as in the previous section. The magnitude of firm i's cost reduction at any
time t is thus given by

$$c_i(\underline{w}) = c_i^0(1 + w_i + \alpha w_j)^{-D_i} \tag{22}$$

where $\underline{w} = (w_1, w_2)$ and c_i^0 is the initial cost of firm i.

The parameter D_i, the rate of innovation, determines the rate at which unit costs decline with accumulated R&D expenditures, from their initial level c_i^0. Under stationary equilibria the accumulated R&D capital w_i will, in the long term, approach a steady state level, say $\hat{w}_i$, implying that the lowest unit cost that a firm can reach by investing in R&D will equal $c_i^0(1 + \hat{w}_i + \alpha \hat{w}_j)^{-D_i}$.

Depending on whether the decisions concerning R&D expenditures are taken independently or jointly by the firms, the following cases will be considered:

(i) Competitive oligopolies and TSCs

We examine first the case of competition in both R&D and in the product market. Each firm independently chooses its output level q_{it} and investment rate u_{it} with the goal of maximizing the present value of its profits over an infinite time horizon, that is

$$\pi_i^k = \sum_{t=0}^{\infty} \rho^t h_i^k(w_{1t}, w_{2t}, q_{1t}, q_{2t}, u_{it}) \tag{23}$$

where

$$
\begin{aligned}
h_i^k(w_1, w_2, q_1, q_2, u_i) = \\
q_i \left[A(q_1 + q_2)^{-\beta} - c_i^0(1 + w_i + \alpha w_j)^{-D_i} \right] - u_i - (1/2)\, \delta_i u_i^2
\end{aligned} \tag{24}
$$

and

$$
h_i^k = \begin{cases} h_i^\sigma, & if \ 0 \le \alpha < 1 \\ h_i^{TSC}, & if \ \alpha = 1 \end{cases} \tag{25}
$$

The superscript σ is employed to denote the case of zero or positive involuntary spillovers, while the superscript TSC implies that the firms - even if they still decide independently on both q_{it} and u_{it} - agree ex-ante to share technological knowledge resulting from R&D activities, though not the costs of these activities. The cost of investment in R&D is given by $u_i + (1/2)\, \delta_i u_i^2$, $\delta_i \ge 0$, where the quadratic term indicates the possibility of diminishing returns to the R&D expenditures.

In the case of a TSC we have that $w_t = w_{1t} + w_{2t}$ with w_{it} $(i = 1, 2)$ evolving according to Eq. (21), which is equivalent to having w_t described by

$$w_{t+1} = (1 - \mu)w_t + u_{1t} + u_{2t} \tag{26}$$

(ii) Research Joint Ventures

Firms taking part in a TSC share technological knowledge but make all production and investment decisions independently of one another. A Research Joint Venture (RJV) goes a step further and, while preserving the independence in setting output levels, it implies cooperation in joint development of R&D projects. Since R&D activities often require substantial investment expenditures, this type of cooperation promises to offer considerable benefits to the industry by eliminating wasteful duplication of research efforts.

In game theoretic terms, a solution to the Competitive duopoly and to the TSC models described above is defined as a non-cooperative Nash equilibrium with respect to both production levels and investment rates. A solution to the RJV model is of a mixed type, i.e., noncooperative with respect to output levels but cooperative with respect to investment rates.

Formally, the RJV model is defined as follows. Two firms compete on the product market with firm i, $i = 1, 2$, seeking to maximize its profit function π_i^{RJV} with respect to q_{it}, $t = 0, 1, \ldots$, where

$$\pi_i^{RJV} = \sum_{t=0}^{\infty} \rho^t h_i^{RJV}(w_t, q_{1t}, q_{2t}) \tag{27}$$

with

$$h_i^{RJV}(w, q_1, q_2, u_i) = q_i \left[A(q_1 + q_2)^{-\beta} - c_i^0 (1 + w)^{-D_i} \right] \tag{28}$$

The state of knowledge, variable w_t, evolves according to

$$w_{t+1} = (1 - \mu)w_t + u_t \tag{29}$$

where u_t denotes the aggregate investment rate in R&D. The sequence of investment rates over the infinite time horizon is determined *jointly* by the two firms as a solution that maximizes their combined profits less the costs of investment, i.e.,

$$\pi^{RJV} = \sum_{t=0}^{\infty} \rho^t [h_1^{RJV}(w_t, q_{1t}, q_{2t}) + h_2^{RJV}(w_t, q_{1t}, q_{2t})] - u_t - (1/2)\delta u_t^2$$

(30)

It is clear that under a RJV the firms need to make a commitment to cooperation and reach an agreement on the division of investment costs that amount to $\sum_{t=0}^{\infty} \rho^t (u_t - (1/2)\delta u_t^2)$. Such a cost sharing agreement can be expected to be straightforward when the firms are similar (the case of a symmetric duopoly) and less so when they are not (the case of an asymmetric duopoly). This issue is addressed in greater detail in Petit and Tolwinski (1999).

4. Markov equilibria

4.1. THE LEARNING-BY-DOING MODELS

The two duopoly models described in section 3 define a two-person nonzero-sum dynamic game with two competing firms as players. The particular equilibrium concept that we shall now consider is based on the following assumptions concerning the way in which the firms determine their output policies.

— Each firm is assumed to maximize its long-term profit as defined by expression (16) subject to the state equation (14) ((20) in the case of a TSC).
— Each firm is assumed to know the value of the state vector $w_t = (w_{1t}, w_{2t})$, i.e., the accumulated outputs of its own and of its competitor, at every time period t.
— Actions of firm i consist in choosing output levels q_{it} on the basis of current information about vector w_t.
— The firms set their outputs independently of each other and do not cooperate in any way.

The above assumptions imply that equilibrium in the duopoly game will be determined in the class of so-called *feedback Nash* or *Markov* equilibrium strategies. The resulting equilibrium has the important property of being *subgame perfect*, that is, at any period t, no matter what happened prior to that period, neither firm can benefit by unilaterally deviating from its equilibrium strategy.

In the case considered here, feedback strategies will be defined as mappings

$$\varphi_i : \mathcal{R}^+ \times \mathcal{R}^+ \longrightarrow \mathcal{R}^+ \tag{31}$$

that associate nonnegative output levels $q_i = \varphi_i(\underline{w})$ with nonnegative values of accumulated production $\underline{w} = (w_1, w_2)$. A pair of feedback strategies $\underline{\varphi}^*(\underline{w}) = (\varphi_1^*(\underline{w}), \varphi_2^*(\underline{w}))$ is said to be a Markov equilibrium if

$$\pi_1^k(\underline{\varphi}^*(\underline{w})) \geq \pi_1^k(\varphi_1(\underline{w}), \varphi_2^*(\underline{w})) \text{ for every } \varphi_1(\underline{w}) \tag{32}$$

and

$$\pi_2^k(\underline{\varphi}^*(\underline{w})) \geq \pi_2^k(\varphi_1^*(\underline{w}), \varphi_2(\underline{w})) \text{ for every } \varphi_2(\underline{w}) \tag{33}$$

where

$$\pi_i^k(\underline{\varphi}(\underline{w})) =$$
$$\sum_{t=0}^{\infty} \rho^t \varphi_i(\underline{w}_t) \left[A(\varphi_1(\underline{w}_t) + \varphi_2(\underline{w}_t))^{-\beta} - c_i^o (1 + w_{it} + \alpha w_{jt})^{-D_i} - c_i^{min} \right]$$
$$\tag{34}$$

for $i = 1, 2$

As we have seen in Sect. 2, a pair of feedback equilibrium strategies $\underline{\varphi}^*(\underline{w}) = (\varphi_1^*(\underline{w}), \varphi_2^*(\underline{w}))$ exists if there exist functions $V_i(\underline{w})$ for $i = 1, 2$ such that the following dynamic programming equations are satisfied.

$$V_1(\underline{w}) = \max_{q_1}\{l_1^k(\underline{w}, q_1, \varphi_2^*(\underline{w})) + \rho V_1(f(\underline{w}, q_1, \varphi_2^*(\underline{w})))\} \tag{35}$$

and

$$V_2(\underline{w}) = \max_{q_2}\{l_2^k(\underline{w}, \varphi_1^*(\underline{w}), q_2) + \rho V_2(f(\underline{w}, \varphi_1^*(\underline{w}), q_2))\} \tag{36}$$

where

$$f(\underline{w}, q_1, q_2) = (w_1 + q_1, w_2 + q_2)^T \tag{37}$$

4.2. THE LEARNING-FROM-R&D MODELS

Consider now how the firms choose their outputs, q_{it}, in the cases of a *Competitive Duopoly* (with *TSC* as a special case) and a *RJV*, respectively.

In both cases the firms compete in the product market and set their outputs independently. Therefore, the problem can be formulated as a nonzero-sum game. Finding Markov equilibria for these models is a little more involved than for the "learning by doing" models, since the decision variables of each player are two in this case. However, since the output variables are absent from the state equations that describe the dynamics of cumulative R&D, output (feedback) strategies for the dynamic game coincide with the Nash equilibrium strategies of the single period games. It is then possible to substitute the output strategies so obtained into the firms' profit functions resulting in new optimization problems, where the objective function of firm i depends only on its investment rate in R&D, u_{it}, and the accumulated technological knowledge w_t.

The procedure therefore consists in two steps. In the first step output equilibrium strategies are obtained by computing feedback Nash equilibria for the single period game. In this case feedback strategies are defined as mappings that associate nonnegative output levels $q_i = \varphi_i(\underline{w})$ with nonnegative values of the accumulated R&D capital $\underline{w} = (w_1, w_2)$. This step leads to the following result.

PROPOSITION 4.1.   A strategy pair $\varphi^*(\underline{w}) = (\varphi_1^*(\underline{w}), \varphi_2^*(\underline{w}))$, where

$$\varphi_i^*(\underline{w}) = \frac{[A(2 - \beta)]^{1/\beta}}{\beta[c_1(\underline{w}) + c_2(\underline{w})]^{\frac{1}{\beta}+1}}[c_j(\underline{w}) - (1 - \beta)\, c_i(\underline{w})] \qquad (38)$$

if

$$c_1(\underline{w}) < P\ (\varphi_1^*(\underline{w}) + \varphi_2^*(\underline{w}))$$
$$c_2(\underline{w}) < P\ (\varphi_1^*(\underline{w}) + \varphi_2^*(\underline{w})) \qquad (39)$$

and

$$\varphi_i^*(\underline{w}) = 0 \qquad (40)$$

$$\varphi_j^*(\underline{w}) = \left[\frac{A(1 - \beta)}{c_j(\underline{w})}\right]^{1/\beta} \qquad (41)$$

if

$$c_i(\underline{w}) > P\ (\varphi_1^*(\underline{w}) + \varphi_2^*(\underline{w})) \qquad (42)$$

with $c_i(\underline{w})$ defined by (22), $(i, j = 1, 2, i \neq j)$ is a feedback Nash equilibrium of the output setting Competitive Duopoly.

Output equilibrium strategies in the RJV case is given by similar expressions with $\underline{w}$ substituted by w and the unit costs defined by $c_i(w) = c_i^0(1+w)^{-D_i}$.

Notice that the feedback strategies defined by Eqs. (38) - (42) are of a stationary type meaning that changes in output over time are due only to changes in the values of the state variables. Also note (Eqs. (40), (41) and (42) that if, at any time t, price becomes lower than firm i's marginal cost, the best reaction for this firm is to stop producing (i.e. leave the market), independently of the level of accumulated knowledge ($\underline{w}_t$). The other firm will then switch to the monopolistic output strategy (Eq.(41)).

In the second step, the equilibrium strategies for setting output levels now obtained can be substituted into the firms' profit functions resulting in new optimization problems which will differ depending on whether the firms choose their investment rates independently (competitive duopoly and TSC) or jointly (RJV).

(i) Competitive Duopoly with TSC as a Special Case

The firms choose their investment rates, u_{it}, independently by maximizing total discounted profits. The new form of firm i's profit function is

$$\pi_i^\sigma = \sum_{t=0}^\infty \rho^t \hat{h}_i^\sigma(w_{1t}, w_{2t}, u_{it}) \tag{43}$$

where

$$\hat{h}_i^\sigma(w_1, w_2, u_i) = \frac{[A(2-\beta)]^{1/\beta}}{\beta[c_1(\underline{w}) + c_2(\underline{w})]^{\frac{1}{\beta}+1}} \frac{[c_j(\underline{w}) - (1-\beta)c_i(\underline{w})]^2}{2-\beta}$$
$$- u_i - (1/2)\,\delta_i u_i^2 \tag{44}$$

if condition (39) holds, while

$$\hat{h}_j^\sigma(w_1, w_2, u_j) = \left[\frac{A(1-\beta)}{c_j(\underline{w})}\right]^{1/\beta}\left[\frac{\beta c_j(\underline{w})}{1-\beta}\right] - u_j - (1/2)\,\delta_i u_j^2 \tag{45}$$

and

$$\hat{h}_i^\sigma(w_1, w_2, u_i) = -u_i - (1/2)\,\delta_i u_i^2 \tag{46}$$

under condition (42) ($i, j = 1, 2$, $i \neq j$). We recall that $c_i(\underline{w}) = c_i^0(1 + w_i + \alpha w_j)^{-D_i}$ where $\alpha \in [0, 1]$. Profit maximization is conducted subject to the state equation (21).

The problem thus defined is a dynamic (Markov) game with investment rates as decision variables. To solve this game we need to compute strategies associating non-negative investment rates with the values of R&D capital $\underline{w} = (w_1, w_2)$. This problem is much harder than the computation of output strategies because the decision variables are present in the definition of the state equation and, consequently, the dynamic game cannot be decomposed into single period games.

A strategy pair $\varphi^*(\underline{w}) = (\varphi_1^*(\underline{w}), \varphi_2^*(\underline{w}))$ is a Markov perfect equilibrium for the dynamic game at hand if there exist functions $V_i(\underline{w})$, for $i = 1, 2$, such that:

$$V_1(\underline{w}) = \max_{u_1}\{\hat{h}_1^\sigma(\underline{w}, u_1,) + \delta V_1(f(\underline{w}, u_1, \varphi_2^*(\underline{w})))\} \qquad (47)$$

and

$$V_2(\underline{w}) = \max_{u_2}\{\hat{h}_2^\sigma(\underline{w}, u_2) + \delta V_2(f(\underline{w}, \varphi_1^*(\underline{w}), u_2))\} \qquad (48)$$

where

$$f(\underline{w}, u_1, u_2) = [(1 - \mu)w_1 + u_1, (1 - \mu)w_2 + u_2]^T \qquad (49)$$

The computation of a Markov equilibrium of the duopoly problem is equivalent to solving the above equations. This is the basis of the policy iteration techniques used for the numerical simulations presented in Sect. 5.

(ii) **RJV**
In the case of a Research Joint Venture, the firms jointly control a single decision variable defining the total R&D expenditures, therefore, the dynamic game becomes a single-agent dynamic optimization problem with objective function

$$\pi^{RJV} = \pi_1^{RJV} + \pi_2^{RJV} \qquad (50)$$

or

$$\pi^{RJV} = \sum_{t=0}^{\infty} \rho^t \hat{h}^{RJV}(w_t, u_t) \qquad (51)$$

where

$$\hat{h}^{RJV}(w,u) = \frac{[A(2-\beta)]^{1/\beta}}{\beta[c_1(\underline{w}) + c_2(\underline{w})]^{\frac{1}{\beta}+1}} \times$$

$$\left\{ \frac{[c_2(\underline{w}) - (1-\beta)c_1(\underline{w})]^2 + [c_1(\underline{w}) - (1-\beta)c_2(\underline{w})]^2}{2-\beta} \right\}$$

$$- u - (1/2)\delta u^2 \qquad (52)$$

if condition (39) holds, or

$$\hat{h}^{RJV}(w,u) = \left[\frac{A(1-\beta)}{c_j(\underline{w})}\right]^{1/\beta} \left[\frac{\beta c_j(\underline{w})}{1-\beta}\right]$$

$$- u - (1/2)\delta u^2 \qquad (53)$$

otherwise $(i,j = 1,2, \ i \neq j)$. The maximization of the profit function is conducted subject to the state equation (29).

5. A numerical example

The dynamic programming equations defined above cannot be solved analytically. However, a computational algorithm based on a policy-iteration method (see the Appendix), makes it possible to find numerical equilibrium solutions.

The models defined in Section 3 have been calibrated (see Petit and Tolwinski, 1996, 1999) and numerical results have been obtained. These results may contribute to the understanding of the behaviour of firms facing the two different forms of learning, and of the impact that the presence of learning can have on the structure of the market.

Only the results concerning the asymmetric case will be discussed here[3], since they provide a good insight into the effects of asymmetries on the evolution of the industrial structure.

5.1. THE CASE OF LEARNING BY DOING

Asymmetries between oligopolistic firms may result from different initial costs, from different rates of learning or from both. As we shall see, in the asymmetric case the decision to share technological information or to keep it proprietary may change not only the market prices and the profit levels of the firms but also the degree of *industrial concentration*. Due to lack of

---

[3] Results concerning the symmetric case can be found in Petit and Towinski (1994, 1996, 1999).

space in what follows we shall consider only cost asymmetries (the results referring to the case of asymmetries in the rate of learning do not differ substantially).

The following values of the parameters have been assumed: $A = 10$, $B = 1.5$, $\rho = 0.90$, $D = 0.20$, $c_i^{min} = 2.0$. The parameter c_i^0 has been modified in the numerical computations, so as to reflect possible asymmetries between the firms. Figure 1 compares output, price and profit trajectories generated by an asymmetric duopoly in the case of different initial unit production costs. More specifically, firm 1 has an initial cost advantage over firm 2: $c_1^o = 2.0$, $c_2^o = 8.0$, $c_1^{min} = c_2^{min} = 2.0$ (i.e. $c_1(0) = 4$ and $c_2(0) = 10$)) though they learn at the same rate $D_1 = D_2 = 0.20$. This is a case of strong asymmetry which clearly shows that the only way for firm 2 to survive in the market is to share technical information with firm 1 (or, in other words, to take advantage of the stronger knowledge of firm 1). In fact, only in this case the production and profit paths of firm 2 are positive at each t.

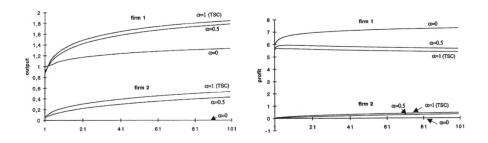

Figure 1. Learning by doing (different initial costs)

As a consequence of the TSC between firm 1 and firm 2, also firm 1 increases its output at each t [4] due to competition in the product market:

---

[4] Except for a very short initial period, due to the higher initial aggressiveness of firms when information is proprietary. See Petit and Tolwinski (1996) for a wider discussion.

the creation of the TSC makes it possible to keep the oligopolistic situation, which will otherwise turn into a monopolistic structure. In this case in fact, if firm 1 keeps its learning proprietary, it will become a monopolist and consequently reduce its production path, making at the same time higher profits. Prices will thus be higher in the case of proprietary information (which turns into a monopolistic structure) than in the case that the two firms decide to share their knowledge (thus keeping the duopolistic structure). From the point of view of consumer welfare, the creation of a TSC is clearly positive, since sharing knowledge between firms will produce higher diffusion of inventions, thus lower costs for both firms, higher production and lower prices.

However, by looking at figure 1, one could wonder why firm 1, the advantaged firm, would accept an agreement implying a transfer of technology to the other firm. Firm 1 knows that in the long run, by holding information proprietary, it will reach a monopolistic situation (and thus higher profits). It is therefore in its own interest to choose this alternative and not the alternative of helping firm 2 to survive. Therefore, in an asymmetric duopoly, the reasons for the firms to create technology cartels seem rather weak. But since there can be high advantages from the point of view of consumer welfare, it should be the concern of the government to create incentives for the stronger firms in order to compensate them from the loss of profits that would derive from the creation of TSCs. These government incentives - unlike what is usually thought - can be an impediment to the formation of monopolistic structures.

5.2. THE CASE OF LEARNING FROM R&D

The results obtained under the assumption of Competition in R&D with and without spillovers are here compared with those obtained in the case in which the firms agree to share technological knowledge (TSC) and with the case in which the firms decide jointly their research expenditures (RJV). In all these cases, we recall, firms act as competitors in the product market. Again, the decision to form a technological cartel (TSC or RJV) may change not only the time trajectories of the relevant variables, but also the *structure of the market*.

As above, asymmetries between oligopolistic firms may result from different initial costs or from different rates of innovation. Again, only the case of cost asymmetries will be considered, since the outcomes in the case of asymmetries in the rate of innovation are substantially similar [5]. As in

---

[5] When asymmetries are very small, the conclusions do not differ substantially from a symmetric case (say, for asymmetries based on values like $c_i^0 = 4.5$ and $c_j^0 = 5$, or $D_i = 0.30$ and $D_j = 0.25$). However, when asymmetries are slightly more significant, the

the case of learning by doing, the creation of technology cartels (both TSC and RJV) are important from a consumer welfare point of view, since such cartels can become the only way to prevent the market from turning into a monopoly.

Figure 2 compares the time trajectories of the output produced by each firm, of total industry output, of price and of the total stock of R&D.

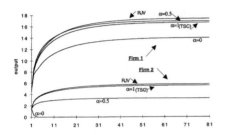
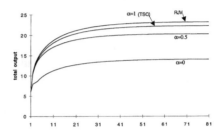
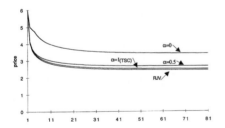
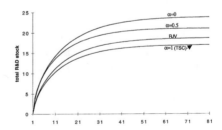

Figure 2. Learning from R&D (different initial costs)

results present very strong differences with those of a symmetric duopoly (see Petit and Tolwinski, 1999).

These trajectories are generated by considering an asymmetric duopoly in the case of different initial unit costs of production. More specifically, firm 1 has an initial cost advantage over firm 2: $c_1^0 = 3.0$, $c_2^0 = 5.0$, though they both have the same rate of innovation $D_1 = D_2 = 0.30$. The values of the remaining parameters are as specified in the previous subsection. This case clearly shows that, in a situation of strong intellectual property rights ($\alpha = 0$), firm 2 will be obliged to stop production and leave the market after three periods. This firm will only survive either in a market with a high degree of spillovers or by making a technological agreement with firm 1. In fact, only in these cases the production path of firm 2 is positive at each t.

As can be seen from Fig. 2, total output is higher and prices are lower for high values of the spillover parameter or when the firms make a technological agreement. In the case of R&D competition, as we have seen, high intellectual protection leads the market towards a monopolistic structure, with only firm 1 producing. Thus, output and prices turn to their monopolistic values. From the consumer's welfare point of view, therefore, the creation of a TSC or a RJV appears to be positive in this case.

Fig. 2 also shows the outcomes obtained for the level of R&D. The total (industry) stock of R&D is higher under a low value of the spillover parameter, since, as mentioned above, for very low values of α, the market becomes a monopoly thus giving rise to a higher level of investment in research [6]. For levels of α close to 0.5 still competition in R&D will produce high levels of research, though, as α increases, the free-riding effect reduces the incentives to invest in R&D. For values of α close to or equal 1 (TSC), firm 2 ceases to invest in research since it can obtain all its knowledge from firm 1 with no costs. Obviously this also reduces the incentives to invest of firm 1 since its investment will help its rival as much as itself (they both have the same rate of innovation D_i). In industries characterized by high values of the spillover parameter ($\alpha \geq 0.8$ in our example), the creation of a RJV will eliminate those free-riding effects and produce a higher stock of research at each t.

Therefore, for markets where technological spillovers are strong, RJVs will produce higher consumer welfare than competition in R&D, since they will give rise to higher levels of research and lower prices. Notice that RJVs produce the lowest path of prices, even lower than that resulting from competition without spillovers. However, as for the case of learning by doing, the advantaged firm should be incentivated to create technological cartels in order to compensate it for a possible loss of profits.

---

[6] The absence of competition in the product market makes it possible for the monopolist to appropriate a higher part of the surplus created by its investments in R&D.

There are two different reasons that explain the welfare benefits of RJVs and TSCs. One concerns *production* advantages. Sharing knowledge makes it possible a quicker unit cost reduction, thus higher output and lower prices. A second reason concerns *industrial concentration*. As we have seen in the case of asymmetric duopolies, technology cartels make it possible to *maintain oligopolistic structures* that would otherwise turn into monopolistic ones, since disadvantaged firms would be obliged to leave the market in the long run. This result may appear surprising at a first sight since, as often underlined in the literature, a *danger* of research cartels is that they may lead to mergers, and therefore to higher industrial concentration. Our results show that this danger may go in the opposite direction.

6. Concluding comments

The main purpose of this paper was to emphasize how a class of dynamic-games (i.e. differential or difference games) can be useful for analysing the process of technological innovation in the framework of an oligopolistic market. Different dynamic oligopoly models have been proposed. Process innovations of the incremental type have been considered, and assumed to be the result of some form of *learning* by the firms: learning by doing and learning arising from R&D. In order to take due account of the relevance of past industrial performance, the models are formulated as Markov games where it is shown how *accumulated* knowledge can, in both cases, be represented by a state variable, the evolution of which is determined by the investment decisions in learning. An investment in learning can be the result of an increase in production (and thus in experience) or an increase in the expenditures in R&D activities.

The approach proposed in this paper also makes it possible to take into consideration the effects of research agreements on the speed of industrial innovation and on the evolution of the industrial structure over time.

Markov perfect equilibria have been computed for the two innovation models considered. Numerical examples are presented which show that research agreements by asymmetric firms can be welfare increasing since, unlike what is often thought, they may prevent industrial concentration.

A drawback of Markov equilibria is the difficulty to derive them analytically (and often also numerically) when the models considered are non-linear (with few exceptions for the linear-quadratic case). The computational algorithm used in this paper makes it possible to find numerical equilibria for a (non-linear) dynamic game with two-players, two state equations and an infinite horizon setting. However it is far from simple to extend the algorithm to the case of more state equations and more players. Research efforts are currently being made in this direction.

APPENDIX

The Numerical Algorithm

An approximation to feedback Nash (Markov) equilibrium for the dynamic games of Section 3 can be found by imposing a finite grid on the state space and then applying a modified policy iteration method (Tolwinski, 1989) to the resulting finite state Markov game. This approach is based on the fact that a strategy pair

$$\underline{\varphi}(\underline{w}) = (\varphi_1(\underline{w}), \varphi_2(\underline{w})) \qquad (54)$$

is a Markov perfect equilibrium for the dynamic game if there exist functions $V_i(\underline{w})$ for $i = 1, 2$ such that the dynamic programming equations (35)-(36) (for the learning-by-doing case) and (47)-(48) (for the R&D case) described in section 4 are satisfied.

These dynamic programming equations can be solved on a finite grid W_h imposed on the state space

$$W = \mathcal{R}^+ \times \mathcal{R}^+ \qquad (55)$$

where

$$W_h = \{(w_i = ih, w_j = jh) : i = 0, \ldots, M; j = 0, \ldots, N\} \qquad (56)$$

with $h > 0$ and

$$M = \text{entier}((w_{1max}/h), \ N = \text{entier}((w_{2max}/h) \qquad (57)$$

The equilibrium strategies $\varphi_i(\underline{w})$ and value functions $V_i(\underline{w})$ are computed only for $\underline{w} = (w_1, w_2) \in W_h$. Notice that W_h is bounded from above by parameters w_{1max} and w_{2max}[7]. The computation of Markov equilibrium strategies for the finite state dynamic game defined on W_h can be carried out by means of the following algorithm.

Algorithm (Modified Policy Iteration)
The following algorithm refers to the "learning-from-R&D" model. It can easily be extended to the "learning-by doing" model[8].

1. Select initial approximations $V_i^0(\underline{w})$ to $V_i(\underline{w})$ for all $\underline{w} \in W_h$, $i = 1, 2$. Set k to 0.

---

[7] In the R&D case, w_{1max} and w_{2max} should be large enough to guarantee that $0 < \hat{w}_1 < w_{1max}$ and $0 < \hat{w}_2 < w_{2max}$, where $(\hat{w}_1, \hat{w}_2)$ is the steady state generated by equilibrium strategies.
[8] The interested reader is referred to Petit and Tolwinski (1996)

2. For every $\underline{w} \in W_h$, compute a Nash equilibrium point, say (u_1^*, u_2^*), of the static game defined by strategy spaces $\mathcal{R}^+$, $\mathcal{R}^+$, and payoff functions $R_i(\underline{w}; u_1, u_2)$, $i = 1, 2$, where

$$R_i(\underline{w}; u_1, u_2) = \hat{h}_i^\sigma(\underline{w}, u_i) + \delta V_i^k(f(\underline{w}, u_1, u_2))$$

Set $\underline{\varphi}^k(\underline{w}) = (\varphi_1^k(\underline{w}), \varphi_2^k(\underline{w})) = (u_1^*, u_2^*)$.

3. Compute approximate values of the payoff functionals corresponding to the strategy pair $\underline{\varphi}^k(\underline{w})$ by applying a few, say $m \geq 1$, Jacobi iterations to the system

$$V_i(\underline{w}) = \hat{h}_i^\sigma(\underline{w}, \varphi_i^k(\underline{w})) + \delta V_i(f(\underline{w}, \underline{\varphi}^k(\underline{w})))$$

for $\underline{w} \in W_h$ and $i = 1, 2$. More precisely, set $V_i^{k,0}(\underline{w}) = V_i^k(\underline{w})$ and for $j = 0, \dots, m-1$ compute

$$V_i^{k,j+1}(\underline{w}) = \hat{h}_i^\sigma(\underline{w}, \varphi_i^k(\underline{w})) + \delta V_i^{k,j}(f(\underline{w}, \underline{\varphi}^k(\underline{w})))$$

Set $V_i^{k+1}(\underline{w})$ to $V_i^{k,m}(\underline{w})$.

4. If $\|V_i^{k+1}(\cdot) - V_i^k(\cdot)\| < \epsilon$ for $i = 1, 2$ then stop; else, set k to $k+1$ and go to step 2.

A computer implementation of the above algorithm requires an interpolation scheme for the calculation of $V_i^k(y)$ for $y \notin W_h$ and a procedure to compute Nash equilibria of static games in step 2. The convergence of the algorithm may depend to considerable extent on the choice of these procedures. One possible approach is to choose an interpolation scheme that generates continuously differentiable value functions $V_i^k(\underline{w})$ and then to use equations

$$\frac{\partial R_i(\underline{w}; u_1, u_2)}{\partial u_i} = 0 \qquad (58)$$

for $i = 1, 2$, to find equilibria in step 2. More specifically, a cubic spline interpolation scheme *in two dimensions* can be used that provides estimates of $V_i^k(y)$ and of its partial derivatives

$$\frac{\partial V_i^k(w_1, w_2)}{\partial w_i} \qquad (59)$$

at $(w_1, w_2) \in W$.

For the problem at hand, the system of equations (58) can be very efficiently solved by a fixed point iteration. Newton's method is another obvious choice. Its value in the given context is limited, however, by the fact that it uses second derivatives of functions $V_i^k(x)$, whose estimates provided by the interpolation tend to be rather inaccurate.

The computational scheme described above turned out to work quite well for the dynamic games considered in this paper. It should be noted, however, that the algorithm is not guaranteed to converge in general and may fail to converge for some combinations of parameters of the model. A major reason for convergence failure is non-uniqueness of equilibrium. However, the model at hand appears to have a unique equilibrium in the class of feedback strictly non-cooperative strategies. This has been verified by running the algorithm for different initial approximations $V_i^0(\underline{w})$ to the value functions. When the algorithm does converge, it generates feedback strategies $\underline{\varphi}$ and value functions V_i, $i = 1, 2$, that satisfy the dynamic programming equations (47) - (48) with some given accuracy ϵ[9].

References

Arrow, K., 1962, "The economic implications of learning by doing", *The Review of Economic Studies*, 29, 155-173.

Basar, T. and G.J.Olsder, *Dynamic Noncooperative Game Theory*, 2nd ed. Academic Press, New York, 1995.

Baumol, W.J., *Entrepreneurship, Management and the Structure of Payoffs*, MIT Press, Cambridge, MA, USA, 1993.

D'Aspremont, C. and Jacquemin, A., 1988, "Cooperative and noncooperative R&D in a duopoly with spillovers" *American Economic Review*, 78, 1133-1138.

Friedman, J.W., *Game Theory with Applications to Economics*, 2nd ed., Oxford University Press, New York,1991.

Kamien, M.I. and N.L. Schwartz, *Market structure and innovation*, Cambridge University Press, Cambridge, UK, 1982.

Katz, M.L. and J.A. Ordover, 1990, "R&D cooperation and competition", *Brookings Papers on Microeconomics*, Brookings Institution, 137-203.

Levin, R.C., A.K. Klevorik, R. Nelson and S.G. Winter, 1987, "Appropriating the results of industrial research and development", *Brookings Papers on Economic Activity*, 3, Brookings Institution, 783-820.

Mansfield, E., 1985, "How rapidly does new industrial technology leak out?" *Journal of Industrial Economics*, 34, 217-223.

Petit, M. L. and B. Tolwinski, "Learning by doing and technology sharing in asymmetric duopolies", in: T. Basar and A. Haurie eds., *Advances in Dynamic Games and Applications*, Birkhauser, Boston, 1994.

Petit, M. L. and B. Tolwinski, 1996, "Technology Sharing Cartels and industrial structure", *International Journal of Industrial Organization*, 15, 77-101.

Petit, M.L. and B. Tolwinski, 1999, "R&D cooperation or competition?", *European Economic Review*, 43, 185-208.

---

[9] In the numerical experiments described in sections 4 and 5, $\epsilon = 10^{-4}$ has been used.

Reinganum,J. F., "The timing of innovation: Research, development, and diffusion" in R. Schmalensee and R. Willig eds., *Handbook of Industrial Organization*, Vol.1, North Holland, Amsterdam, 1990.

Selten, R., 1975, "Reexamination of the perfectness concept for equilibrium points in extensive form games, International Journal of Game Theory", 4, 25-55.

Tolwinski, B., "Newton-type methods for stochastic games", in T. Basar and P. Bernhard eds. *Differential Games and Applications*, Springer-Verlag, Heidelberg, 1989.

CHAPTER 14

THE STRUCTURE OF FAIR-DIVISION PROBLEMS AND THE DESIGN OF FAIR-NEGOTIATION PROCEDURES

MATTHIAS G. RAITH (`mraith@wiwi.uni-bielefeld.de`)
Institute of Mathematical Economics
University of Bielefeld
P.O. Box 100131
D-33501 Bielefeld
Germany

Abstract. In this paper we analyze the structure of bilateral fair-division problems and show how it carries over to a wider spectrum of multi-issue negotiations. This provides insight into the extension of fair-division algorithms to more general fair-negotiation procedures. We develop a simple procedural tool that can be used to implement a variety of formal bargaining-theoretic solution concepts, where implementation here is supposed to mean the actual realization of a cooperative outcome in practice. An important feature is that the algorithm requires only little computational effort. Due to its tractability, it thus also provides the necessary argumentative support for a cooperative negotiation process.

1. Introduction

Negotiation analysis is a practical field of research in which the objective is to develop theoretical approaches that will help us understand and resolve actual conflicts. Although it shares a common ground with formal bargaining theory, one typically finds that bargaining theorists and negotiation analysts live and work in different worlds. According to Brams and Taylor (1996), "bargaining theory has proved singularly inapplicable to the settlement of real-life disputes" because of its "divorce" from theories of fair-division.[1] By viewing negotiation problems as problems of fair-division, they show how practical procedures can be applied to complex conflicts. A fair-division problem has all the characteristics of a bargaining problem.

---

[1] Brams and Taylor (1996), p. 66.

There are, however, two additional elements that make fair-division models particularly attractive for a prescriptive approach to negotiations.

First, there is an apparent notion of fairness involved, meaning that norms play an important role in finding an agreement. As Young (1991) argues, norms help negotiators to coordinate their expectations, because plausible standards narrow the range of acceptable and, therefore, credible agreements. Moreover, fairness principles also help to enhance the durability of agreements, because the latter are based on standards that are relatively stable over time. Norms in fair division play the same role as the solution concepts of cooperative game or bargaining theory. A solution, derived within a normative framework and characterized by a bundle of desirable properties (the axioms), is a *norm*. The fair-division model thus approaches a problem with a problem-solving solution concept.

The second characteristic feature of the fair-division approach is that the fair outcome is determined by a procedure rather than a formula. It may seem artificial, at first, to consider cooperative solutions in connection with procedures, because in cooperative game theory neither the definition of a game nor the specification of a solution includes the plan of play. From this perspective, a procedure appears to be superfluous. It is important to note, though, that the procedure is not intended to complement the formula, but rather designed to substitute for it in the form of an algorithm.

The psychological advantage of a procedure is that it allows players to actually participate in the implementation of a solution. Often negotiation support systems fail to acknowledge this point. A computer program that merely calculates bargaining solutions is not always useful, because players want to understand the solution as well the process that leads them there – hitting the 'return' button alone does not provide much insight. In addition, it is important for players to actively take part in the process, i.e. they want to play the game themselves. Closely related, but from a decision-analytic perspective, is the argumentative aspect: A procedure decomposes the implementation of a solution into successive steps. If the steps are intuitive, they will support the communication process by providing the arguments which, according to Keeney and Raiffa (1976), are necessary to implement a solution.

In this paper we show how the structure of bilateral fair-division problems can provide insight into the extension of fair-division procedures to more general multiple-issue negotiations. We formulate a division problem involving multiple items as a multiple-issue negotiation, where each issue has two discrete options that determine which player receives the item. We focus on the algorithm 'Adjusted Winner,' introduced by Brams and Taylor (1996), which is designed to implement an efficient and equitable outcome, i.e., where parties enjoy equal percentage gains.

By showing explicitly how Adjusted Winner uses the structural information contained in the multi-issue negotiation, we are able to identify and modify the main elements that allow for its generalization. This enables us to develop a simple procedural tool that can be used to implement a variety of formal solution concepts, where implementation here is supposed to mean the actual realization of a cooperative outcome in practice. An important feature is that the algorithm requires little computational effort. Due to its tractability, it thus also provides argumentative support for a cooperative negotiation process.

In section 2, we first discuss and analyze the structure of simple fair-division problems that involve the division of an estate. In section 3, we then describe the algorithm Adjusted Winner. We extend this fair-division algorithm in section 4 to a fair-negotiation procedure that can be applied to multiple-issue multiple-option negotiations. We conclude in section 5 and point to some extensions of our analysis.

2. The Structure of a Fair-Division Problem

2.1. A GRAPHICAL CHARACTERIZATION

Consider the problem of dividing an estate consisting of 6 items between two parties M and R, and assume that players have additively separable preferences over the items. This allows us to characterize preferences by normalized utilities that sum to one (hundred). The valuations of player M and R are given in Table I.

TABLE I.   Valuations of an estate

	1	2	3	4	5	6	Σ
M	25	8.$\bar{3}$	8.$\bar{3}$	8.$\bar{3}$	25	25	100
R	10	10	10	40	10	20	100

According to Table I, for each player the value of the estate is equal to 100%. The point allocations for the individual items can be viewed as players' interests in these items measured by their percentage share of the whole estate. For example, the percentage shares could be derived from subjective monetary valuations. A common monetary standard is not necessary, though, because players can also use the aggregate of all items (i.e. the whole estate) as a standard of value. This is particularly interesting in situations where a monetary valuation of items is not appropriate. In that case, it may be much simpler for parties to allocate 100 percentage points among the items to be divided. Brams and Taylor (1999) show how

to use this point allocation scheme for a variety of different problems such
as divorce settlements, territorial conflicts, or political disputes, where a
monetary valuation often is not possible.

At first sight, the assumption of additive separability is quite strong.
For example, if items 2, 3, and 4 are the three volumes of an encyclopedia,
which M values at 25% of the estate, then it is implausible that a single
volume by itself has 1/3 of the value – in other words, dividing an encyclopedia presumably lets its value decrease more than proportionately. The
value of an individual volume would then depend on how many volumes
M receives altogether.[2] In order to achieve additive separability, Keeney
and Raiffa (1991) suggest regrouping items until this condition is satisfied.
Although this may be difficult in practice, there are apparently enough
real-world problems that fit (at least approximately) into this structure to
also consider it theoretically.

If individual items are not divided, then there are $2^6 = 64$ possible
allocations. For each of the 6 items there are two possibilities: Either player
M receives the item or player R gets it. Players' total utilities $u^x = u_1^x + \cdots + u_6^x$, $x = M, R$, for each of the 64 agreements are indicated by the hollow
points in Figure 1. Implicit in this characterization is the assumption that,
if a player does not receive an item $i = 1, \dots, 6$, then the player's subutility,
u_i^x, is zero.

If all items are divided into two separate packages, and if a player values
one package at x points, then the other package must have a value of
$(100-x)$ points for her. Equivalently, if one package is valued at $x = (50+z)$
points, then its counter-package has a value of $(50 - z)$ points. Since this
applies for both players, the outcomes in Figure 1 are point-symmetric
around the 50-50 allocation. Take, for example, the allocation $\{4, 5, 6\}$ for
M and $\{1, 2, 3\}$ for R, yielding $u_A = (58.\bar{3}, 30)$ (point A). If both players
now exchange their packages, then the new outcome is $u_{A'} = (41.\bar{6}, 70)$
(point A'). Other examples shown in Figure 1 are the symmetric locations
of points C and C' or D and D'. The 50% point thus divides the normalized
utility space into four separate regions.

Of particular interest is the north-east region, because this shows allocations where both players receive packages of items providing each with more
than 50% of the "pie." Hence, neither player will wish to trade packages.
This is not the case in the other three regions, where at least one player
is envious of the package that the other one has. The allocations in the
north-east region are thus 'envy-free.'

---

[2] That player R values the volumes asymmetrically is not such a problem. For instance,
her valuation of the complete encyclopedia may be 60%, but mainly because of the
excellent coverage of, say Zambia, which is in the third volume.

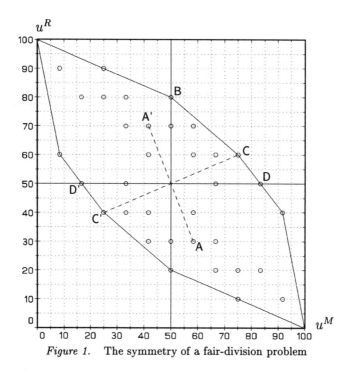

Figure 1.   The symmetry of a fair-division problem

Next, consider the efficient outcomes of this fair-division problem. If we impose the assumption that players have linear preferences over subdivisions of items, then the subutilities can be characterized by linear functions $u_i^x : [0,1] \rightarrow I\!R$, $x = M, R$. The assumption of divisible items allows us to theoretically acknowledge convex combinations of packages.[3] This reduces the number of discrete (undivided) efficient packages, since only those agreements will be considered as efficient that are not dominated by convex combinations of other packages. Raiffa (1996) refers to these agreements as 'extreme-efficient' outcomes. In Figure 1, the efficiency boundary is indicated by the solid piecewise linear curve that intersects the north-east quadrant. Due to the point symmetry of outcomes, the Pareto curve has a pendant running through the south-west quadrant, thus providing a lower boundary for fair-division outcomes.

In order to move from one discrete outcome to the next on the efficiency curve, only one item must be shifted from one player to the other. This shift does not require, nor can it be achieved through the trading of items.

---

[3] Although most items are physically divisible (King Solomon even threatened to divide a child), they may lose much of their value, as we argued above with the encyclopedia. Hence, finding procedures that minimize the division of individual items is of particular interest.

For an illustrative proof, we use Raiffa's (1996) graphical derivation of the efficiency curve. We begin by giving all items to one player, say player R, yielding the outcome $u = (0, 100)$. At this point, a movement to the next discrete outcome on the efficiency frontier can only be achieved by passing one item from R to M, since M has nothing to trade. For an efficient shift of items, the loss of player R relative to the gain of player M must be minimized. The lowest cost-gain ratio is achieved with items 1 and 5. Moving first one item and then the other to player M leaves M with the package $\{1, 5\}$ and R with $\{2, 3, 4, 6\}$, leading to the outcome B in Figure 1. The next efficient shift involves item 6, where R loses 20 and M gains 25; this brings players to point C. Note that, up to this point, there is no trade involving the exchange of more than one item that can achieve a lower cost-gain ratio. Assume that there is such a trade: Then M must give an item, say item 5, back to R in return for item 6. This leaves M with the package $\{1, 6\}$, implying that the movement to this package yields a lower cost-gain ratio than the movement to package $\{1, 5\}$, which is a contradiction. This line of reasoning applies to any point along the efficiency boundary: If an efficient transfer of one item were to require the return of a previously shifted item, then it cannot have been efficient to shift the previous item in the first place.

The efficient movement along the Pareto frontier by transferring one item at a time has an important practical implication for fair-division procedures: Any efficient division of items, i.e. any allocation on the efficiency boundary, requires the actual division of at most one single good. As a consequence, any procedure based on the division of more than one item, e.g. a settlement where every item is split equally (50-50), will generally lead to an inefficient outcome.[4] How much efficiency is sacrificed clearly depends on the location of the efficiency curve. The question, therefore, is: What determines the shape of the fair-division problem?

2.2. CONFLICTING VS. DIVERGING INTERESTS

Consider a modified version of the problem in Table I, where parties' preferences over the individual items are now characterized by the entries in Table II. The notable difference is that now players' interests are much more similar than before. Indeed, there are even items that are valued the

---

[4] As one can see in Figure 1, this applies to other splits as well. According to Young (1996), in the quick partitioning of India and Pakistan in 1947, much of the government property was split 80:20, which approximately represented the relative sizes of the populations. This is, in fact, an example of a division where dictionaries from the national libraries were split A-K, L-Z.

same by both players. The joint utilities for each of the 64 agreements are shown in Figure 2.

TABLE II.   Conflicting valuations of an estate

	1	2	3	4	5	6	Σ
M	25	10	5	15	20	25	100
R	20	15	5	20	20	20	100

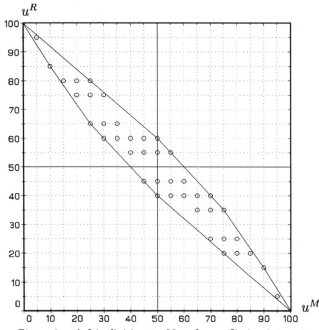

Figure 2.   A fair-division problem for conflicting interests

Since the convergence of interests narrows the field of outcomes around the negative diagonal, it is more difficult for players to both achieve more than 50% of the pie. Thus, the conflict potential of this fair division problem is higher than before, because players gain mainly at the expense of each other. This is the classical picture of bargaining as a "tug-of-war," e.g. between a buyer and a seller haggling over the price of a good. Indeed, when there is only one item to be divided, then the outcomes will all lie on a negatively sloped 45° line. This is because both players value the item at 100 points, so that bargaining becomes a constant-sum game. With multiple items, the more players converge in their interests, the more the aggregate over all items becomes one big item, which both players desire for themselves. Conflicts arise when interests converge, i.e. when both players want the same things with the same intensities.

If similar interests increase the conflict between players, then diverging interests should have the opposite effect. Assume, therefore, that players differ strongly in their interests – what one player desires the other barely cares for. An example is given in Table III. In this case, if players cooperate and distribute items according to interests, then they should both be able to obtain a sizable share of the pie. For each of the 64 possible agreements, the joint outcomes are shown in Figure 3.

TABLE III.   Diverging valuations of an estate

	1	2	3	4	5	6	Σ
M	25	5	10	10	25	25	100
R	5	20	5	60	5	5	100

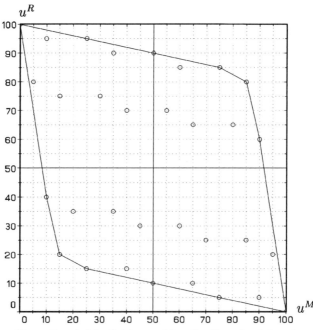

Figure 3.   A fair-division problem for diverging interests

It would surely be surprising if a situation with such a wide range of outcomes created a conflict between parties. Indeed, Figure 3 illustrates that both players can achieve more than 80% of the pie. For example, if items 1, 5, and 6 are given to M, items 2 and 4 are given to R, and item 3 is split 2:1 between M and R, then both players can achieve an outcome of 81.$\bar{6}$%.

As we have seen above, the negotiation over a single item is purely distributive. Thus, the simplest way to widen the range of outcomes in north-east direction is to increase the number of items on the table. In an actual fair-division problem this would require that the parties themselves bring additional items to the table. Of course, this will only work if parties' interests over the newly included items differ. Young (1996) argues that another method for widening the problem is to view the conflictual items (those that are valued the same by both players) as a bundle of attributes, which are valued differently. By 'unbundling attributes' players can both enjoy higher gains.[5]

Even with a problem as wide as in Figure 3, it is not clear that parties will succeed in finding a favorable solution. With only six items it should be fairly easy to achieve a mutually beneficial outcome. However, considering that n items allow 2^n outcomes, the problem becomes increasingly complex. A wide range of outcomes thus creates considerably more room for joint gains, but the efficiency frontier may become difficult to find without analytic support.

3. Adjusted Winner

3.1. THE BASIC ALGORITHM

Brams and Taylor (1996) introduce an algorithm 'Adjusted Winner,' which is designed to implement an efficient and equitable outcome, where equitability implies that both players achieve the same percentage gains, under the assumption that players do not strategically misrepresent their preferences. Our graphical characterization of fair-division problems in the previous section illustrates an important result for two-person fair-division problems, namely that an equitable, efficient outcome is always envy-free. As Raith and Welzel (1998) show, Adjusted Winner, in its basic form, implements the same outcome that is induced by the axiomatic solution of Kalai and Smorodinsky (1975). The algorithm is remarkably simple and consists of only two steps: The first step implements an efficient outcome,

---

[5] A classic story in the negotiation literature tells of two sisters attempting to divide an orange fairly. By simply cutting it in half, each would receive 50%. However, when asked about their preferences, one says that she wants only the juice for drinking, and the other replies that she wants only the peel for baking. Hence, by appropriately dividing the orange, both sisters can achieve 100% of their interests

and the second step adjusts this outcome until it is equitable. Table IV summarizes the procedure.

TABLE IV.   Adjusted Winner

	M	R	cost/gain
Item 1	$\boxed{25}$	10	$25/10 = 2.5$
Item 2	$8.\overline{3}$	$\boxed{10}$	
Item 3	$8.\overline{3}$	$\boxed{10}$	
Item 4	$8.\overline{3}$	$\boxed{40}$	
Item 5	$\boxed{25}$	10	$25/10 = 2.5$
Item 6	$\boxed{25}$	20	$25/20 = 1.25$
Estate Value	100	100	
Step 1	75	60	M → R
Transfer	-25	+20	Item 6
	50	80	M ← R

Equitability: $50 + \alpha 25 = 60 + (1 - \alpha)20$
M's Share of Item 1: $\alpha = 2/3$

% of Estate	$66.\overline{6}$	$66.\overline{6}$

In the first step of Adjusted Winner, players individually submit sealed bids for each of the items to be distributed. The bidding procedure ensures that players' preferences are revealed simultaneously. In the most general form, players' bids are percentage values representing their relative interests in the individual items.[6] For our example, the bids are as in Table I.

The bidding procedure is resolved by giving each item to the highest bidder, i.e. to the player who values this item most. If players value an item equally, then the item can be allocated in any form, e.g. by tossing a coin. For our example of Table I, step 1 assigns items 1, 5, and 6 to player M, while player R receives items 2, 3, and 4. The resulting utilities are $u^M = 75$ and $u^R = 60$.

The intriguing feature of step 1 is that it guarantees efficiency. In order to see this, one must only verify that there is no trade of items that can benefit both players. If each item is given to the player who values it most, then this distribution also maximizes the sum of players' utilities – remember the assumption that the other player's utility of not receiving the item is zero. Consequently, if one player gains by trading items, the other must lose, since trading cannot increase the sum of both players' utilities.

---

[6] If all items can be valued in some common unit of account, say Dollars, then the conversion to percentage terms is not necessary.

There is also no convex combination of item distributions that dominates the allocation of step 1, since then at least one of these distributions must yield a higher sum of players utilities.[7]

Step 2 of Adjusted Winner now shifts items from one player to the other until equitability is achieved. With the outcome of step 1 already efficient, step 2 is designed to preserve this feature by transferring items at the lowest cost-gain ratio, i.e. by moving along the efficiency frontier. In our example, utility gains must be shifted from player M to player R. The only items to consider are those that have been given to M. Items 1 and 5 are valued by M at 25 percent and by R at 10 percent. A transfer would thus imply that M loses 2.5 times as much as R gains. Item 6 is valued at 25 percent by M and 20 percent by R, implying a cost-gain ratio of only 1.25 (cf. Table IV). Hence, the efficient adjustment shifts item 6 first. Note that a complete transfer of this item leads to $u^M = 50$ and $u^R = 80$, which would require a transfer back from R to M. Equitability thus requires a division of item 6, given by some fraction $\alpha \in [0, 1]$, such that

$$50 + \alpha 25 = 60 + (1 - \alpha)20 \ .$$

The equitable split is given by $\alpha = 2/3$, i.e., M must give up $1/3$ of her possession of item 6. The resulting outcome is $u^{M*} = u^{R*} = 66.\overline{6}$.

3.2. STRATEGIC BEHAVIOR

Our analysis of Adjusted Winner is based on the assumption of what Raiffa (1996) calls 'partially open truthful exchange.' This is when parties act according to their preferences, but they are only reluctant to reveal them. The situations that we have in mind are typical for ongoing relationships in business, political, and social life. Parties wish to play cooperatively and play "truthfully" according to the rules, but they simply do not trust each other enough to lay down their cards. In a symmetric situation, both players know only their own preferences. With a bidding procedure such as Adjusted Winner, players are forced to reveal their preferences simultaneously. Since the procedure features a timing that rules out exploitation afterwards, there is no room for strategic behavior.

The situation is notably different with asymmetric information, i.e. when one player, say M, has complete information, and the other, R, knows only her own preferences. In order to illustrate this, consider once more the problem of dividing an estate consisting of six items, where players' *true* preferences are given in Table I on page 207. Assume that player M has complete information, while R plays truthfully, knowing only her own valuation of the estate. Suppose that both wish to play by the rules of Adjusted

---

[7] cf. Brams and Taylor (1996): Proof of Theorem 4 1.

Winner. The only point at which M can play strategically is during the bidding process, because afterwards the procedure "takes over" until the fair outcome is implemented. If R plays truthfully, Adjusted Winner will guarantee her a share of at least 50% – either by assigning her an item or through compensation with an item. Player M, knowing this, can then play strategically by lowering her bid on the items that she values higher than R, and raising her bid on the items that she values lower than R, while preserving the ordinal relationship between both players' valuations. An example is shown in Table V, where the valuations for R are the same as in Table I.

TABLE V.   Strategic valuations of an estate

	1	2	3	4	5	6	$\sum$
M	13	9	9	35	12	22	100
R	10	10	10	40	10	20	100

Table V shows the strategic bids of player M, while Table I on page 207 contains her true valuations. The *true* and the *strategic* fair-division problems are illustrated together in Figure 4. The hollow points characterize the true problem, which is identical to that of Figure 1 (p. 209). The solid points indicate the strategic problem. Observe that player R's valuations are the same for both situations – only M's valuations change the shape of the problem.

Figure 4 also illustrates the goal of strategic misrepresentation of preferences under a fair-division procedure. Since Adjusted Winner implements an equitable outcome, i.e. on the 45°-line, player M must strategically "adjust" her preferences as closely as possible to those of player R. With strongly conflicting interests, player R cannot expect to achieve much more than 50%. For the example shown in Table V, Adjusted Winner assigns items 1, 5, and 6 to player M yielding a value of 47, while items 2, 3, and 4 go to player R who receives 60. The temporary outcome is denoted by point T in Figure 4. The adjustment from R to M now requires the division of either item 2 or 3 – since $9/10 > 35/40 = 7/8$ – such that player M receives a share of 68.42% of item 2, for example. This yields a seemingly equitable outcome of 53.16%, denoted by point A in Figure 4. While this outcome holds true for player R, player M's true valuation, according to Table I, is 80.70%. This actual outcome is denoted by point B in Figure 4.

Note, however, that M must be careful in "adjusting" her preferences. For example, if she bids too high for item 4, say 37, and too low for the other two items, say 8 for each, then item 4 will be chosen for division,

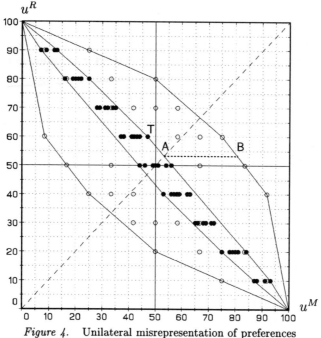

Figure 4.   Unilateral misrepresentation of preferences

instead of 2 or 3. With player R's valuation of item 4 four times as high as her valuation of 2 or 3, the *true* adjustment will be more costly for player M. Hence, strategic manipulation of fair-division procedures, although theoretically possible, can be quite demanding in practice, when several items are involved. If a player's knowledge of the other's preferences is not fully reliable, bidding too close to the other player can be dangerous. Indeed, player M can then easily end up with an item she does not want, because she now has to compensate R for her loss.

It is important to note that Adjusted Winner enhances the difficulty of strategic manipulation. With the restriction of the 100-point allocation, strategic bidding on only one item is not possible. Thus if M strategically raises her bid on a specific item, she must lower her bid on at least one other item. As Brams and Taylor (1996) argue, in order to take advantage of the other player one must, therefore, consider several issues simultaneously. Consequently, Adjusted Winner is de facto more robust against manipulation than procedures where players can bid strategically for single issues.

With symmetric full information, i.e. when the parties both know each others' preferences, a fair-negotiation procedure can only be a useful support tool in implementing a fair and efficient outcome, if both parties are engaged in 'full open truthful exchange'.

If non-cooperative players use Adjusted Winner strategically, then this
will result in an equilibrium allocation with a 50-50 split. The challenge
for the procedural designer is then to construct mechanisms that induce
rational players to implement the desired outcome through their strategic
interaction. One must not forget, though, that both players must agree
to such a mechanism. With full rationality they then know what the outcome will be. Hence, committing oneself to this mechanism is the same
as committing oneself to the induced solution. The same applies to any
mechanism under complete information. Just because players are behaving
strategically, this does not imply that there is less commitment involved.

4. Multiple Issues with Multiple Options

The fair-division problems that we have considered so far can be viewed as
specific types of bargaining problems. Each item to be distributed can be
regarded as a negotiated issue in a multiple-issue negotiation, where there
are two discrete options for resolving each issue: option 1, player M receives
the item and player R receives nothing; or option 2, player R receives the
item and player M receives nothing. In addition, we have considered convex
combinations of these two extreme winner-take-all outcomes.

In general, however, negotiation goes beyond the mere distribution of
items. Indeed, there is hardly any restriction for the issues on an agenda,
and only the bounds on negotiators' creativity limit the number of options
for resolving the issues at stake. Even for problems of fair division there are
often more than just the two winner-take-all outcomes. Various methods
are discussed by Young (1996). For example, each player may wish an
item for herself, but both would rather have no one receive it (which is
a third option) than let the other one have it for herself. This method of
"subtraction" is sometimes used for the resolution of territorial disputes.

Consider the following example of a multiple-issue multiple-option negotiation taken from Gupta (1989). A manufacturer M and a retailer R
negotiate over six alternative marketing plans (the options) for each of six
different products (the issues). Their interests in the individual plans are
characterized by the percentage gains shown in Table VI. An agreement
consists of one option for each of the six issues. The total percentage gain
that a player realizes is obtained by adding the points of the individual
chosen options.

This method of weighting issues and valuing options is quite common
in negotiation analysis. Players first weight issues by distributing 100 percentage points among the issues to be negotiated, thus characterizing their
interests. They then value options by giving the best option of an issue

TABLE VI.   Percentage gains of players M and R from plans A–F for products 1–6

ISSUES

		1	2	3	4	5	6	Σ
A	M	0.00	0.00	8.3$\bar{3}$	0.00	25.00	0.00	
	R	10.00	10.00	0.00	40.00	0.00	20.00	
B	M	6.87	3.00	7.08	4.17	20.8$\bar{3}$	5.8$\bar{3}$	
	R	8.25	7.50	3.00	33.35	5.00	17.00	
C	M	12.50	5.00	6.25	5.21	15.25	11.6$\bar{6}$	
	R	7.00	6.00	5.00	25.00	6.10	14.00	
D	M	15.25	6.25	5.00	6.25	12.50	15.63	
	R	6.10	5.00	6.00	20.00	7.00	12.50	
E	M	20.8$\bar{3}$	7.08	3.00	7.50	6.87	20.8$\bar{3}$	
	R	5.00	3.00	7.50	8.60	8.25	10.00	
F	M	25.00	8.3$\bar{3}$	0.00	8.3$\bar{3}$	0.00	25.00	
	R	0.00	0.00	10.00	0.00	10.00	0.00	
Max	M	25.00	8.3$\bar{3}$	8.3$\bar{3}$	8.3$\bar{3}$	25.00	25.00	100.00
	R	10.00	10.00	10.00	40.00	10.00	20.00	100.00

(left margin vertical label: OPTIONS)

the same number of points that are assigned to this issue, while the worst option is valued at 0 points. Intermediate options are valued in between. In this example, the percentage gains are derived from the net profits that players can obtain from a specific marketing plan. Note that for each player the minimum percentage gain over all six issues is 0 and the maximum 100. With six possible options to each of the six issues, there are $6^6 = 46,656$ possible agreements. The joint outcomes for both players are plotted in Figure 5.

Clearly, this problem has a greater degree of complexity than the fair-division problems considered in the previous sections. Indeed, as Gupta (1989) verified in experiments with this specific problem, players (in real life) do have difficulty in finding an efficient agreement.

If one were to approach this negotiation problem with a fair-division procedure, one would view the issues as items. As a consequence, only the weights of the issues would be acknowledged but not the values of all the options. In particular, the winner-take-all assumption implies that only the extreme options (A and F) are taken into account, so that either player M or player R achieves her maximum percentage gain, while the other player receives nothing. Note that the two bottom rows of Table VI reproduce the fair-division problem of Table I.

With negotiation focused on only two options to each issue, there are

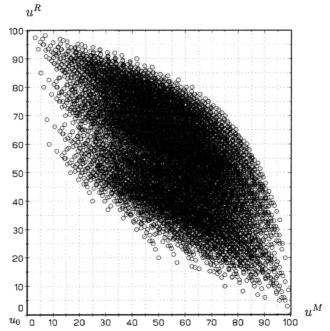

Figure 5.   A marketing negotiation with 46,656 possible agreements

again 64 possible agreements. Applying Adjusted Winner then yields the same allocation as in the previous section. The outcome is denoted by point AW in Figure 6. The only difference is that now the agreement is over the realization of marketing plans rather than the distribution of items.

For comparison, Figure 6 also shows the Pareto frontier of the full negotiation problem illustrated in Figure 5. As one can clearly see, it lies well beyond the Pareto frontier of the fair-division problem. The reason is that the marketing negotiation includes a number of efficient intermediate options that are precluded when players concentrate on only the extreme options. For each individual issue the loss of efficiency may not be very large, but summed over all issues, there is substantial value left on the table.

The challenging question is thus whether simple fair-division procedures can be extended to deal with more complex problems that involve several options. As Raith and Welzel (1998) show, the necessary modification of Adjusted Winner is, in fact, straight-forward once the steps are formulated more generally.

In step 1, Adjusted Winner implements an efficient allocation without reference to any particular notion of fairness. This is achieved by simply maximizing the sum of players' utilities for each individual issue. If this

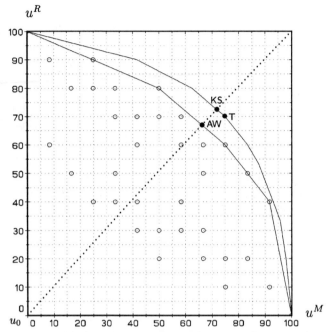

Figure 6.   Outcomes of Adjusted Winner for the (two-option) fair-division problem and the (multiple-option) negotiation problem

approach is applied to the problem of Table VI, then the outcome is not the six-tuple of extreme options [FAFAAF], i.e. where, for each issue, the option is chosen that benefits the player who values this issue highest. Instead, the agreement that maximizes the sum of players' utilities is [EDCABE], which yields the individual outcomes $u^M = 75$ and $u^R = 70$. This temporary agreement is denoted by point T in Figure 6.

In step 2 of Adjusted Winner, the objective is to find an equitable outcome without having to give up efficiency. At point T this requires shifting gains from player M to R at the lowest cost-gain ratio. As we recall from our analysis of fair-division problems, transfers are accomplished by shifting one item at a time. In particular, there is no trading of items involved. In the context of multiple-option issues, an efficient shift now implies changing one option at a time, i.e., there is no trading of options across issues.

The comparison of all issues reveals that issues 2 and 3 both offer efficient transfer opportunities at a lower cost-gain ratio than any of the other issues.[8] As one can quickly check, the agreement [EC(0.9$\overline{4}$D+0.0$\overline{5}$F)ABE]

---

[8] Issue 4, of course, allows no further transfer at all, since player R already has her most preferred option.

gives each player an efficient and equitable percentage gain of 72.$\overline{2}$%. This outcome, denoted by point KS in Figure 6, characterizes the Kalai-Smorodinsky solution for the multiple-issue multiple-option marketing negotiation. Note, again, that the final agreement involves only one convex combination of two options of a single issue.

The implementation of an efficient and equitable solution is achieved through an intuitive generalization of the steps of Adjusted Winner. As our example illustrates, the required calculations are remarkably simple. Indeed, computer support may actually be counter-productive, since it tends to obscure the individual steps of the algorithm. This is a crucial aspect, because an implementation algorithm can only offer procedural support if it provides the argumentative basis that is required to motivate and justify the solution.

5. Conclusions

Our objective in this paper was to convey a theoretical understanding of the structure of multiple-issue negotiation problems, which can be used to develop practical procedures that assist negotiators in finding a cooperative agreement. Starting with problems of fair division, we showed that these feature structural elements that carry over to more general negotiation problems. Likewise, there are procedures designed for this simple structure that can be extended to more general fair-negotiation procedures. A procedure designed to implement an efficient and fair outcome is particularly useful for negotiation problems, where parties have strongly diverging interests in the issues at stake, because then there are significant opportunities for joint gains. As our analysis demonstrates, with a supporting procedure, players can actively expand the pie by increasing the number of issues, without having to worry about the increase in complexity

By reformulating the fair-division procedure 'Adjusted Winner,' we found a practical algorithm that can be used to implement cooperative bargaining solutions in complex multiple-issue negotiations. In our example, we focused on the implementation of the Kalai-Smorodinsky solution, but Adjusted Winner can easily be adapted to implement other solution concepts as well. All that is required is a modification of step 2 of the algorithm, which determines the final point on the efficiency frontier. Moreover, it is straight forward to acknowledge further important aspects such as the influence of outside options or reference points on the negotiated outcome.

The link between practical procedures and formal theoretical concepts opens a door for further research in negotiation analysis, since it enables a formal theoretical foundation for developing procedural support tools that can be employed in negotiation practice.

Acknowledgements

I would like to thank an anonymous referee for helpful comments. Financial support by the 'Ministerium für Wissenschaft und Forschung, Nordrhein-Westfalen,' is gratefully acknowledged.

References

Brams, S.J., Taylor, A.D.: 1996, *Fair Division – from cake-cutting to dispute resolution*, Cambridge, MA: Cambridge University Press

Brams, S.J., Taylor, A.D.: 1999, *The Win/Win Solution: Equalizing Fair Shares to Everybody*, New York: W.W. Norton (forthcoming)

Gupta, S.: 1989, 'Modeling Integrative, Multiple Issue Bargaining,' *Management Science* **35**, pp. 788–806

Kalai, E., Smorodinsky, M.: 1975, 'Other Solutions to Nash's Bargaining Problem,' *Econometrica* **43**, pp. 513–518

Keeney, R.L., Raiffa, H.: 1976, *Decisions with Multiple Objectives*, New York: Wiley

Keeney, R.L., Raiffa, H.: 1991, 'Structuring and Analyzing Values for Multiple-Issue Negotiations,' in H.P. Young (ed.): *Negotiation Analysis*, Ann Arbor: University of Michigan Press

Raiffa, H.: 1996, *Lectures on Negotiation Analysis*, Cambridge, MA: The Program on Negotiation

Raith, M.G., Welzel, A.: 1998, 'Adjusted Winner: An Algorithm for Implementing Bargaining Solutions in Multi-Issue Negotiations,' *University of Bielefeld, Institute of Mathematical Economics,* WP 295

Young, H.P.: 1991, 'Fair Division,' in H.P. Young (ed.): *Negotiation Analysis*, Ann Arbor: University of Michigan Press, pp. 25–45

Young, H.P.: 1996, 'Dividing the Indivisible,' in Zeckhauser, R.J, Keeney, R.L., and Sebenius, J.K. (eds.): *Wise Choices – Decisions, Games, and Negotiations*, Boston: Harvard Business School Press

CHAPTER 15

EFFECTIVITY FUNCTIONS AND PARLIAMENTARY GOVERNANCE STRUCTURES

STEFANO VANNUCCI (`vannucci@unisi.it`)
Universita' di Siena,
Dipartimento di Economia Politica,
P.za S.Francesco 7, 53100 Siena, Italia

Abstract. Several parliamentary governance structures based upon a directly elected premier are analyzed through their effectivity functions.It is shown that only collegial governance structures which provide a tight connection between the premier and her prefixed majority enjoy strong stability.

1. Introduction

Parliamentary systems operating in multiparty environments typically suffer from an instability threat. This fact might at first appear to be an inescapable reality. Indeed, parliamentary systems may be properly *defined* by their reliance on executive-removal procedures based upon *non-confidence* majority votes. But then, whenever the executive is supported by a *coalition* majority — a quite likely occurrence in a multiparty environment — instability is and has to be a very real possibility, or so it seems. On the other hand, presidential systems have ostensibly their own serious drawbacks (involving essentially the possibility of weak, divided governments). Thus, the general stability issue for parliamentary systems is apparently a most significant one. It remains to be seen whether "stabilizing" rules can be devised in order to effectively address this problem. As a matter of fact a typically proposed remedy is *direct election of the premier*. From a general game-theoretic perspective this proposal makes good sense because it amounts to introducing a focal player that might indeed help to shrink down the set of effective coalitions (see e.g. Peleg (1981) for a general approach to coalition formation based upon *dominant focal players*). Thus, it should not come as a surprise that direct election of the premier is the focus of recent enacted or proposed institutional reforms both in Israel and Italy. Of course, several variants of this move can be distinguished, depending on whether the premier and her majority-coalition are jointly elected or not, and according to the details of executive-removal and legislature-termination procedures. Several substantive question immediately arise. Can some version of directly elected premiership work as a stabilizing device in a multiparty parliamentary system? If so, which one(s)? And why should they work?

The present paper tentatively addresses those issues by analyzing several parliamentary governance structures with a directly elected premier. This is done by attaching *effectivity functions* —i.e. *abstract models of coalitional power allocation* — to such governance structures, and studying the stability properties of the former. It is shown that different versions of premiership have indeed significant consequences with respect to stability of outcomes, but most of them fail. In particular, it turns out that a *tight connection between the elected premier and a prefixed majority coalition is crucial if generic stability is to be obtained.* The paper is organized as follows: section 2 is devoted to a short intuitive description of a few stylized parliamentary structures with a directly elected premier; section 3 includes the proposed effectivity-function-model of governance structures, and the results; a few concluding remarks are offered in section 4. The proofs are confined to an appendix.

2. Governance Structures

Clearly enough, the stability of a governance structure depends both on its institutional rules and on the characteristics of the relevant actors (e.g. the strength of the party system, and the nature of the underlying conflicts of interest). The present paper, however, will focus on the role of *institutional rules.* I consider the following general scenario. A premier is selected through general elections. The premier (de facto) appoints the executive that will remain in charge until the legislature issues a majority non-confidence vote against it. Thus, the setting is essentially a *parliamentary* one. Also, in what follows it is implicitly assumed that the executive can call for a confidence vote at any moment. However, this basic scenario admits several variants. I shall single out for analysis the following list of stylized governance structures.

• G_0 (*Constructive non-confidence*) : *The premier and the executive are subject to a (non-)confidence vote that consists in the appointment of a substitute premier.*

G_0 is a version of "constructive non-confidence" procedures that are currently used in Germany and Spain, as combined with direct election of the premier. I will represent G_0 by a family $E_0(.)$ of effectivity functions — to be defined in the next section— that endow any majority with full power concerning premiership and the executive (this fact will be reflected by *simplicity* of $E_0(.)$ i.e. its equivalence of a simple game as endowed with an outcome set). Indeed, G_0 rules out the occurrence of premier-vacancy, but it merely discourages executive-removal and majority change. Thus, it is unlikely to prevent —by itself— even cabinet instability, and that commonsensical prediction is confirmed by the analysis of $E_0(.)$ provided in the next section.

- G_1 (*Legislature-terminating Non-confidence*) : *The premier and the executive are subject to a (non-)confidence vote. In case a majority non-confidence vote is issued -and only in this case- new elections are held.*

G_1 embodies a "rigid" deterrent against instability, namely automatic legislature-termination. This governance structure will be therefore represented by a parameterized effectivity function $E_1(.)$ such that the premier has veto power on every policy except that any majority can force new elections.

- G_2 (*Semi-Presidential Premiership without Legislature-terminating Power*): *The executive is subject to a (non-)confidence vote, while the premier is not. New elections are held at a pre-fixed date.*

G_2 retains (semi-)parliamentary features in that the executive- if not the premier- is subject to a (non-)confidence vote. In the next section it will be represented by a parameterized effectivity function $E_2(.)$ where both the premier and any majority can veto any outcome, but none of them can force new elections. Indeed, a governance structure such as G_2 is essentially isomorphic to a presidential systems with some degree of power-sharing between executive and legislature.

- G_3 (*Semi-Presidential Premiership with Costly Legislature-terminating Power*): *The executive is subject to a (non-)confidence vote, while the premier is not. The premier can call for new elections by resigning.*

- G_3' (*Semi-Presidential Premiership with Free Legislature-terminating Power*): *The executive is subject to a (non-)confidence vote, while the premier is not. The premier can call for new elections without having to resign.*

Clearly enough, G_3 and G_3' typically differ with respect to the costs of (and incentives for) legislature-termination on the part of the premier, but not in terms of the stricto sensu power allocation among players and coalitions. Thus, they will be both represented by a parameterized effectivity function $E_3(.)$ where both the premier and any majority can veto any outcome, but the premier can also call for new elections.

- G_4 (*Symmetric Unilateral Terminating Power*) : *The premier and the executive are subject to a (non-)confidence vote. New elections are held if either a majority issues a non-confidence vote or the premier resigns, and asks for them.*

G_4 endows both the executive and the legislature with the right to call for new elections. Obviously, G_4 will be represented by a parametrized effectivity function $E_4(.)$ where any majority, and the premier alone, enjoy both the power to veto any outcome *and* the power to call for new elections.

• G_5 (*Joint Power with Variable Majority*) : *The premier and the executive are subject to a (non-)confidence vote. New elections are held if a majority issues a non-confidence vote and the premier does not resign.*

Under G_5, new elections require a consensus of sort between the executive and the legislature (a version of this arrangement is currently adopted in Sweden, where the appointed premier can call for new elections after a non-confidence vote). G_5 will be represented by a parameterized effectivity function $E_5(.)$ such that both the premier and any majority can veto every outcome, and even new elections can only be forced jointly by them.

• G_6 (*Collegial Power with a Fixed Majority*) : *The premier and the executive are subject to a (non-)confidence by their electorally fixed majority. New elections are held if either that majority breaks down or the premier resigns.*

Under G_6 the premier and her/his majority form a collegium. Therefore, G_6 is to be represented by a parameterized effectivity function $E_6(., M)$ — where M is the elected collegial-majority— such that each member of the collegium can veto any outcome, and all of them can jointly enforce every single outcome.

As mentioned above, in the next section the stability properties of the foregoing governance structures will be scrutinized by attaching to each of them a suitable parameterized family of effectivity functions, i.e. a family of coalitional game forms that represent the allocation of a priori decision power among coalitions over the relevant outcome space.

3. Model and Results

Let N, X be two non-empty sets (denoting the player set and the outcome set, respectively: non-empty subsets of N denote *coalitions*, non-empty subsets of X denote *issues*). An *effectivity function (EF)* on (N, X) is a function $E : P(N) \to P(P(X))$ (where $P(Y)$ denotes the power set of Y) s.t.:

$$EF(i) : E(N) = P(X)\backslash \{\emptyset\} \,;$$
$$EF(ii) : E(\emptyset) = \emptyset;$$
$$EF(iii) : X \in E(S) \text{ for any } S \in P(N)\backslash \{\emptyset\} \,;$$
$$EF(iv) : \emptyset \notin E(S) \text{ for any } S \in P(N)\backslash \{\emptyset\} \,.$$

A few structural properties of EFs will be used in the sequel, namely

Regularity: An EF E on (N,X) is *regular* if $A \cap B = \oslash$ entails either $A \notin E(S)$ or $B \notin E(N/S)$;

Maximality: An EF E on (N,X) is *maximal* if $A \notin E(S)$ entails $B \in E(N\backslash S)$ for some B s.t. $A \cap B = \oslash$;

Monotonicity: An EF E on (N,X) is *monotonic* iff it satisfies both *N-monotonicity* (i.e. for any A, S, T: $S \subseteq T$ and $A \in E(S)$ entail $A \in E(T)$) and *X-monotonicity* (i.e. for any S, A, B: $A \subseteq B$ and $A \in E(S)$ entail $B \in E(S)$);

Superadditivity: An EF E on (N,X) is *superadditive* iff for any $S, T \subseteq N$, $A, B \subseteq X$:
if $A \in E(S), B \in E(T)$ and $S \cap T = \emptyset$ then $A \cap B \in E(S \cup T)$;

Simplicity: An EF on (N,X) is *simple* if a set $W \subseteq P(N)$ exists s.t. $A \in E(S)$ iff ((either $A = X$ and $M \neq \emptyset$) or ($A \neq \emptyset$ and $S \in W$)).

REMARK 1. *It should be noticed that superadditivity entails regularity and N-monotonicity, and that regularity and monotonicity must be satisfied if E is to be interpreted as a representation of the outcomes that coalitions can guarantee to themselves by a suitable coordination of their players' action according to some strategic game form. (Moreover, superadditivity and X-monotonicity suffice to warrant that interpretation of E: see e.g. Otten, Borm, Storcken, Tijs (1995)). Given our aims, this is probably the most adequate formulation of the notion of "forcing" power of coalitions, but certainly not the only one. Another possibility consists in focussing on the power of coalitions to "counteract" actions on the part of their opponents. This leads us to the notion of a* polarity *operator for EFs (see Abdou, Keiding (1991)).*

DEFINITION 2. *(Polar of an EF)* Let (N,X) be a pair of non-empty sets and E an EF on (N,X). Then the *polar* E^* of E is the EF on (N,X) that is defined by the following rule:
for any $S \subseteq N$, $S \neq \emptyset$:
$E^*(S) = \{B \subseteq X : B \cap C \neq \emptyset \text{ for any } C \in E(N\backslash S)\}$, and $E^*(\emptyset) = \emptyset$.

It is easily shown that, for any EF E on a pair (N,X), monotonicity of E entails monotonicity of E^*, and $E \subseteq E^*$ iff E is regular, whereas $E^* \subseteq$

E iff E is maximal (see e.g. Abdou, Keiding (1991)). Those facts jointly imply that whenever an EF E is *both regular and maximal its stability properties (to be discussed below) are invariant with respect to the proposed interpretation of coalitional a priori decision power.*

As usual, *coalitional stability* of an EF E will be assessed by looking at its core correspondence on a suitable domain of preference profiles : in particular we shall focus on *strong stability* of such a correspondence, a requirement that embodies an "optimistic" version of Von Neumann-Morgenstern stability (see e.g. Peleg (1984), Demange (1987), Abdou, Keiding (1991)).

DEFINITION 3. *(Core Stability) Let (N, X) be a pair of non-empty sets, E an EF on (N, X), and $D(N, X)$ a domain of preference profiles $\succ =$ $(\succ_i)_{i \in N}$, where — for any $i \in N$ — $\succ_i$ is a suitable binary relation on X. The core of E at $\succ$ —written $C(E, \succ)$— is the set of $(E, \succ)$-undominated outcomes i.e.:*

$$C(E, \succ) = \{x \in X : \text{ for no } S \subseteq N, B \subseteq X, B \in E(S)$$
$$\text{and } b \succ_i x \text{ for any } b \in B, i \in S\}$$

An EF E is (core-)stable *on $D(N, X)$ if $C(E, \succ) \neq \emptyset$ for any $\succ \in$ $D(N, X)$, and* unstable *otherwise.*

In this paper two domains of preference profiles will be considered for any pair (N, X), namely the set $D^0(N, X)$ of all $(N-)$profiles of weak orders on X (i.e. asymmetric and negatively transitive binary relations) that have a maximal element, and the set $D^1(N, X)$ of all $(N-)$profiles of Euclidean preferences on X with a bliss point (in this case we posit $X \subseteq R^k$ for some positive k, and for any $i \in N$, $x, y \in X$, $x \succ_i y$ iff $d(x, z_i) < d(y, z_i)$, where d denotes the Euclidean metric and z_i is the bliss point of player i).

Unfortunately, even if $C(E, \succ) \neq \emptyset$ an $(E, \succ)$-dominated outcome may well turn out to be dominated solely through subsets of X having no core elements. Since preferences are typically *not* verifiable, this opens up the possibility of coalitional manipulation of outcomes (see e.g. Demange (1987)). That fact motivates the following strengthening of the core-stability requirement :

DEFINITION 4. (Strong Stability of an EF) *Let E be an EF on (N, X) and $D = D(N, X)$ a domain of preference profiles as defined above. Then, E is said to be* strongly (core-)stable *if for any $\succ \in D(N, X)$ the following holds true: for any $x \in X \backslash C(E, \succ)$, a pair $S \subseteq N, B \subseteq X$ exists such that $B \cap C(E, \succ) \neq \emptyset$, $B \in E(S)$ and $b \succ_i x$ for any $b \in B$, $i \in S$.*

The following notions and properties are crucial when scrutinizing core-stability properties of an EF:

DEFINITION 5. (Lower Cycles of an EF) *A lower cycle of an EF E for (N, X) is a finite sequence $((S_i, B_i))i = 1, .., h)$ such that $\cup_{i=1}^{h} B_i = X$, $B_i \in E(S_i), i = 1, .., h$, $\cap_{i=1}^{h} S_i = \emptyset$, and $B_i \cap B_j = \emptyset$ for any $i, j = 1, ..., h$, $i \neq j$. (see Abdou, Keiding (1991)).*

DEFINITION 6. (Cycles of an EF) *A cycle of an EF E for (N, X) on domain D^* is a finite sequence $((S_i, B_i), i = 1, .., h)$ with $B_i \in E(S_i)$ for any $i \in N$, and such that another sequence $((C_i), i = 1, .., h)$ which satisfies the following properties exists:*

i) $\emptyset \neq C_i \subseteq X$, $\cup_{i=1}^{h} C_i \supseteq X$, $C_i \cap C_j = \emptyset$, $i, j = 1, .., h$, $i \neq j$;

ii) for any $i \in \cup_{j=1}^{h} S_j$ a preference relation $\succ_i$ for i in D^* exists such that $b \succ_i c$ for any $k \in \{j : i \in S_j\}$, and any $(b, c) \in B_k \times C_k$ (which implies $B_k \cap C_k = \emptyset$ whenever $\succ_i$ is asymmetric, whence w.l.o.g. $B_i \neq X$).

REMARK 7. *The explicitly domain-dependent definition of cycle given above is due to Kolpin (1991). An essentially equivalent alternative definition is due to Abdou, Keiding (1991). It can be easily shown that, for any domain D^* of preference profiles, a lower cycle is also a cycle on D^*.*

DEFINITION 8. (Acyclicity) *An EF E on (N, X) is acyclic on domain $D^* = D^*(N, X)$ if there are no cycles of E on D^*.*

DEFINITION 9. (Convexity) *An EF E on (N, X) is convex if $B_i \in E(S_i), i = 1, 2$ entails either $B_1 \cap B_2 \in E(S_1 \cup S_2)$ or $B_1 \cup B_2 \in E(S_1 \cap S_2)$.*

It is well-known that *acyclicity* of an EF E on a given domain D^* (of acyclic preference profiles) is a *necessary and sufficient condition for core-stability of E on D^** (see Abdou, Keiding (1991), Kolpin (1991)). Hence, in particular, *absence of lower cycles is a necessary condition* for core-stability of an EF E on a domain D^*. Moreover, *convexity* —which implies super-additivity and is in turn jointly implied by acyclicity and maximality—is a *sufficient condition for strong stability* of an EF on a given domain of acyclic preference profiles (see Demange (1987), Abdou, Keiding (1991)).

I turn now to the *representation of governance structures by means of effectivity functions. Let player 1 denote the directly elected premier and $N = \{1, ..n\}, n \geq 3$ the set of parties* controlling seats within a h-sized legislature as characterized by their respective weights (i.e. seats) $w_1, .., w_n$. Thus, the weight profile $w = (w_1, .., w_n)$ is such that $\sum_{i \in N} w_i = h$, where w_1 denotes the number of seats controlled by the premier's party (we allow for $w_1 = 0$), and $h \geq n$ (the notation $w(S) = \sum_{i \in S} w_i$ for any $S \subseteq N$ will also be used). Such a w will be denoted as a *weight profile for a h-sized legislature on N*, or a h-sized weight profile on N. The outcome set X —with $\#X \geq 3$— represents the space of policies

under the executive's jurisdiction plus a   *"deadlock"* outcome y_0 whose
exact interpretation should be suitably adapted according to the governance
structure under consideration (e.g. y_0 may denote "no government" or "no
action"—hence "status quo"—), and a *"restart"* state x_0 denoting "new
elections". Thus X will henceforth denote —unless the contrary is explicitly
stated (see the scheme $E_2(.)$ as defined below)— a set with *two distinguished
elements (i.e.outcomes x_0, y_0)* and such that $\#X \geq 3$.We shall also say that
an ordered pair (N, X) of non-empty sets is $h-admissible$ if $h \geq \#N - 1$,and
non-trivial if both $h \geq \#N$ and $\#X \geq 3$ hold. As usual, simple majority
voting is assumed. Therefore, the winning —i.e. all powerful—coalitions are
exactly those coalitions which include the premier —and her party—and
control at least $\lfloor h/2 \rfloor + 1$ seats. In order to avoid irrelevant complications I
shall focus on the class of legislatures characterized by *strong* weight profiles
(or seat allocations) as defined below.

DEFINITION 10. *A weight profile w for a h-sized legislature on N is said
to be strong if for any $S \subseteq N$ either $w(S) \geq \lfloor h/2 \rfloor + 1$ or $w(N \backslash S) \geq
\lfloor h/2 \rfloor + 1$.*

NOTATION 11. *The set of strong weight profiles for a h-sized legislature
on N will be denoted by $W^S(h, N)$.*

I proceed now to define, for any non-trivial pair (N, X), a list of w-
parameterized EFs on (N, X) which I claim can be most plausibly attached
to the parliamentary governance structures informally outlined above.

NOTATION 12. *Let us consider a fixed positive integer $h \geq 3$. A parameterized family $E(., (N, X); h)$ of EFs on a given h-admissible, non-trivial
pair (N, X) —with parameter set $V((N, X); h)$ —is said to enjoy property P on $V((N, X); h)$ if and only if $E(v, (N, X); h)$ satisfies P for all
$v \in V((N, X); h)$.*

DEFINITION 13. *Let $h \geq 3$ be a positive integer, (N, X) an h-admissible,
non-trivial pair, $V(N, X; h)$ a suitably defined parameter set. An $h-$sized
EF-scheme $E(.; h)$—with parameter scheme $V(.; h)$— is the collection of
all such $V((N, X); h)$-parameterized families as (N, X) varies among h-
admissible, non-trivial pairs. A (dimension-free) EF-scheme E(.) —with
parameter scheme $V(.; .)$ —is the set of all such h-sized EF-schemes (with
h a positive integer, $h \geq 3$).*

NOTATION 14. *An h-sized EF-scheme $E(.; h)$ will be said to satisfy an
EF- property P from the list introduced above if every EF which belongs
to some of its parameterized families satisfies P. A (dimension-free) EF*

*scheme will be said to satisfy such an EF-property P if — for any $h \geq$
3— $E(.; h)$ satisfies P. In particular, when referring to stability properties
it will be said that the h−sized EF-scheme $E(.; h)$ is (strongly) stable on
domain $(V(.; h), D^*(.; h))$ iff $E(v, (N, X); h)$ is (strongly) stable on $D^* =
D^*(N, X)$ for all $v \in V((N, X); h)$ and all h-admissible non-trivial pairs
(N, X).Similarly a (dimension free) EF-scheme will be said to be (strongly)
stable on domain $(V(.), D^*(.))$ iff for any $h \geq 3$ the h-sized EF-scheme
$E(.; h)$ is (strongly) stable on domain $(V(.; h), D^*(.; h))$. An EF-scheme that
is not stable on a certain domain will also be said to be unstable on that
domain.*

REMARK 15. *We are interested in those legislatures which are "large
enough" to be of practical interest (hence the restriction to the case $h \geq 3$).
Moreover, for any given legislature, we regard the number of parties as
essentially indeterminate from the institutional design perspective of the
present paper. This perspective is also the reason we focus on EF-schemes,
and insist on stability at each profile for each EF of the EF-scheme under
consideration..*

DEFINITION 16. (The EF-scheme $E_0(.)$ for Constructive Non-Confidence).
*Let $h \geq 3$ be the size of the legislature, (N, X) a h-admissible, non-trivial
pair, and w a weight profile for a h-sized legislature on N.
Then $E_0(w, (N, X); h)$ is the EF for (N, X) defined as follows: for any
$S \subseteq N, B \subseteq X$:*

$$B \in E_0(w, (N, X); h)(S) \text{ iff}$$
$$\text{either } (B \neq \emptyset \text{ and } w(S) \geq \lfloor h/2 \rfloor + 1) \text{ or } (B = X, S \neq \emptyset)$$

Thus, under $E_0(.; h)$ any majority coalition is fully empowered, or winning. Indeed, the "constructivity constraint" on non-confidence votes does
not affect the allocation of a priori decision power among coalitions. The
following result on $E_0(.; h)$, which is mentioned here for the sake of comparisons, is a trivial corollary of well-known results concerning core-instability
of majority —or more generally non-weak—simple games (see e.g. Schofield
(1985)).

PROPOSITION 17. *$E_0(.)$ is simple, monotonic, superadditive, maximal,
and unstable on $(W^S(,), D^1(.))$.*

DEFINITION 18. (The EF-scheme $E_1(.)$ for Legislature-terminating Non-
Confidence) *Let $h \geq 3$ be an integer—the size of the legislature—, (N, X)
a h-admissible, non-trivial pair, and w be a weight profile for a h-sized
legislature on N. Then, $E_1(w, (N, X); h)$ is the EF on (N, X) defined as*

follows: for any $S \subseteq N$, $B \subseteq X$,
$B \in E_1(w,(N,X);h)(S)$ *iff* $(x_0 \in B, w(S) \geq \lfloor h/2 \rfloor + 1)$, *or* $(B \neq \emptyset, w(S) \geq \lfloor h/2 \rfloor + 1)$, *or* $(B = X, S \neq \emptyset)$.

Under $E_1(.)$, only majorities which include the elected premier are fully empowered, while any majority is able to force new elections. Clearly enough, $E_1(.)$ is *not* simple, hence results on simple games do not apply here. Nevertheless, the following instability result obtains:

PROPOSITION 19. *i)* $E_1(.)$ *is monotonic, superadditive,* not *maximal, and* unstable *on* $(W^S(.), D^0(.))$;
ii) $E_1^*(.)$ *is also* unstable *on* $(W^S(.), D^0(.))$.

DEFINITION 20. (The EF-scheme $E_2(.)$ for Semi-Presidential Premiership without Legislature-terminating Power) *Let $h \geq 3$ be an integer, X a set with* one *distinguished element y_0 (the "deadlock" outcome), and such that $\#X \geq 3$, (N,X) a h-admissible, non-trivial pair, w a weight profile for a h-sized legislature on N . Then, $E_2(w,(N,X);h)$ is the EF on (N,X) defined as follows: for any $S \subseteq N$, $B \subseteq X$, $B \in E_2(w,(N,X);h)(S)$ iff $(y_0 \in B, w(S) \geq \lfloor h/2 \rfloor + 1)$ or $(B \neq \emptyset, w(S) \geq \lfloor h/2 \rfloor + 1$, and $1 \in S)$ or $(B = X , S \neq \emptyset)$.*

Thus, under $E_2(.)$ majorities that include the premier are fully empowered, while majorities that do *not* include the latter can only block any action (but cannot call for new elections).

PROPOSITION 21.   *i)* $E_2(.)$ *is monotonic, superadditive,* not *maximal, and* unstable *on* $(W^S(.), D_-^0(.))$;
ii) $E_2^*(.)$ *is also* unstable *on* $(W^S(.), D^0(.))$.

DEFINITION 22. (The EF-scheme for Semi-Presidential Premiership with Legislature-terminating Power). *Let $h \geq 3$ be an integer, (N,X) a h-admissible non-trivial pair, w a weight profile for a h-sized legislature on N. Then $E_3(w,(N,X);h)$ is the EF on (N,X) defined as follows: for any $S \subseteq N, B \subseteq X$, $B \in E_3(w,(N,X);h)(S)$ iff $(\{x_0,y_0\} \subseteq B, w(S) \geq \lfloor h/2 \rfloor + 1)$ or $(x_0 \in B, 1 \in S)$ or $(B \neq \emptyset, w(S) \geq \lfloor h/2 \rfloor + 1, 1 \in S)$ or $(B = X, S \neq \emptyset)$.*

Thus, $E_3(.)$ allows the premier to call for new elections, and any majority to block any decision, apart from new elections. Only those majorities that include the premier are endowed with full decision power

PROPOSITION 23. *i)* $E_3(.)$ *is monotonic, superadditive,* not *maximal, and* unstable *on* $(W^S(.), D^0(.))$;
ii) $E_3^*(.)$ *is also* unstable *on* $(W^S(.), D^0(.))$.

DEFINITION 24. (The EF-scheme $E_4(.)$ for Symmetric Unilateral Terminating Power). *Let $h \geq 3$ be an integer, (N, X) a h-admissible non-trivial pair, w be a weight profile for a h-sized legislature on N. Then, $E_4(w, (N, X); h)$ is the EF on (N, X) defined as follows : for any $S \subseteq N, B \subseteq X$,*
$B \in E_4(w, (N, X); h)(S)$ *iff* $(x_0 \in B, 1 \in S)$ *or* $(x_0 \in B, w(S) \geq \lfloor h/2 \rfloor + 1)$
or $(B \neq \emptyset, 1 \in S, w(S) \geq \lfloor h/2 \rfloor + 1)$ *or* $(B = X, S \neq \emptyset)$.

Under $E_4(.)$ both the premier and any majority are empowered the current legislature and premiership and call (unilaterally) for new elections. As usual, those majorities which include the premier are the only fully empowered coalitions.

PROPOSITION 25. $E_4(.)$ *is monotonic, superadditive, maximal and unstable on* $(W^S(.), D^0(.))$.

DEFINITION 26. (The EF-family $E_5(.)$ for Joint Power with a Variable Majority). *Let $h \geq 3$ be an integer, (N, X) a h-admissible non-trivial pair, w a weight profile for a h-sized legislature on N. Then, $E_5(w, (N, X), h)$ is the EF on (N, X) defined as follows: for any $S \subseteq N, B \subseteq X$,*
$B \in E_5(w, (N, X); h)(S)$ *iff* $(B \neq \emptyset, w(S) \geq \lfloor h/2 \rfloor + 1, 1 \in S)$, *or* $(x_0 \in B, \#B \geq 2, 1 \in S)$ *or* $(B = X, S \neq \emptyset)$.

It follows from the foregoing definition that — under $E_5(.)$ — the premier is able to enforce any outcome *provided that a majority does not block her decisions* (e.g. with a non-confidence vote), in which case new elections are to be held.

PROPOSITION 27. *i) $E_5(.)$ is monotonic, superadditive, not maximal, and stable (but not strongly stable) on $(W^S(.), D^0(.))$;*
ii) $E^*(.)$ is unstable on $(W^S(.), D^0(.))$.

DEFINITION 28. (The EF-scheme $E_6(.)$ for Collegial Power with a Fixed Majority) *Let $h \geq 3$ be a positive integer, (N, X) a h-admissible non-trivial pair, w a weight profile for a h-sized legislature on N, and $M \subseteq N$ a coalition such that $1 \in M$ and $w(M) \geq \lfloor h/2 \rfloor + 1$. Then $E_6(w, (N, X), M; h)$ is the EF on (N, X) defined as follows: for any $S \subseteq N, B \subseteq X$, $B \in E_6(w, (N, X), M; h)$ iff $(B \neq \emptyset, S \supseteq M)$ or $(x_0 \in B, S \cap M \neq \emptyset)$*

Clearly enough, under $E_6(.)$ a *fixed majority M that includes the premier* is the only winning —i.e. fully empowered—coalition. It follows that each member of M has a veto right : exercising such a veto right results in new elections.

PROPOSITION 29. $E_6(.)$ *is monotonic, maximal and convex, hence superadditive and* strongly stable *on* $(W^S(.), D^0(.))$.

Therefore, $E_5(.)$ and $E_6(.)$ are the only EF-schemes in our list that enjoy
stability, and $E_6(.)$ is the only strongly stable one. Intuitively, this result
can perhaps be ascribed to the fact that $E_5(.)$ enjoys the *mild stabilizing
power of a unique veto player* (who may lack, however, a majority support),
while $E_6(.)$ *can rely on the strong stabilizing power of a genuinely collegial
governance structure.*

4. Concluding Remarks

The results presented in this paper suggest that a direct election of the premier is far from being sufficient to ensure outcome stability in a multiparty
parliamentary system: a lot of supplementary rigidity has to be injected
into the system if the instability threat is to be ruled out by institutional
design. At the same time, our results suggest that if stability in the outcome
space —as opposed to mere cabinet stability—is the ultimate concern, then
the sort of rigidity induced by presidential-like governance structures is not
the "right" kind of rigidity (see the proposition on the $E_2(.)$ scheme as
presented above). Rather, *a tight connection between the premier and her
prefixed majority is required for outcome stability.*

5. Appendix: Proofs

Proof of Proposition 17. Simplicity and monotonicity of $E_0(.)$ follow trivially from its definition, while maximality follows immediately from the definition of a strong weight profile. Superadditivity is easily checked, noticing
that —by definition of $E_0(.)$ and $W^S(.,.)$ — for any integer $h \geq 3$, h-
admissible non-trivial pair (N, X) and h-sized $[A \in E_0(w, (N, X); h))(S)$,
$B \in E_0(w, (N, X); h)(T)$ and $S \cap T = \emptyset]$ entails $A \cup B \in \{A, B\}$ whence
$A \cap B \in \{A, B\}$. To check instability just take $\#N \geq 3, h = 3, w = (1, 1, 1)$,
and bliss points z_1, z_2, z_3 with $z_i \neq z_j$, $i, j = 1, 2, 3$. Then, instability
follows from a standard geometric argument on the 2-simplex spanned by
$\{z_1, z_2, z_3\}$.   ∎

Proof of Proposition 19. i) Monotonicity follows trivially from the definition of $E_1(.)$, and superadditivity is also straightforward (indeed, for
any relevant $h, (N, X)$, and $w \in W^S(h, N)$, $A \in E_1(w, (N, X); h)$ and $B \in
E_1(w, (N, X); h)$ entail $w(S) \geq \lfloor h/2 \rfloor + 1$, $w(T) \geq \lfloor h/2 \rfloor + 1$, hence $S \cap T \neq
\emptyset$). Clearly, $E_1(.)$ is not maximal, hence $E_1(.) \neq E_1^*(.)$: to see this, take
an integer $h \geq 3$, a h-admissible non-trivial pair of sets (N, X), and a h-
sized strong weight profile $w \in W^S(h, N)$. Next, take $x \in X, S \subseteq N$ s.t.

$x \neq x_0, 1 \notin S$, and $w(S) \geq \lfloor h/2 \rfloor + 1$. Then, $\{x\} \notin E_1(w, (N, X); h)(S)$ and $B \notin E_1(w, (N, X); h)(N \backslash S)$ for any $B \subseteq X \backslash \{x\}$.

In order to check for instability, just consider the following example: $\#N = 4$, $h = 3$, $w = (0, 1, 1, 1)$, $X = \{x_0, x_1, x_2\}$ (here, we assume $y_0 \in \{x_1, x_2\}$).

It is immediately checked that the following sequence of coalition-issue pairs:

$$((S_1 = \{1, 2, 3\}, B_1 = \{x_1\}),$$
$$(S_2 = \{1, 2, 4\}, B_2 = \{x_2\}), (S_3 = (\{3, 4\}, B_3 = \{x_0\})$$

is a lower cycle of $E_1(w, (N, X); h)$. Then, $C(E_1(w, (N, X); h), \succ) = \emptyset$, if one takes $\succ$ as follows :

$$((x_1 \succ_1 x_2 \succ_1 x_0), (x_1 \succ_2 x_2 \succ_2 x_0), (x_0 \succ_3 x_1 \succ_3 x_2), (x_2 \succ_4 x_0 \succ_4 x_1)).$$

ii) Observe that, by definition of the core correspondence, for any pair (N, X), and any pair E, E' of EFs on (N, X) such that $E \subseteq E'$, i.e. $E(S) \subseteq E'(S)$ for each $S \subseteq N$:

$$C(E', \succ) \subseteq C(E, \succ) \text{ at any relevant preference profile } \succ .$$

But then, regularity of $E_1(.)$ —which is implied by superadditivity and implies $E_1(.) \subseteq E_1^*(.)$, as mentioned above— entails instability of $E_1^*(.)$, since $E_1(.)$ itself has been shown to be unstable. ∎

Proof of Proposition 21. Same as the proof of Proposition 19 above, provided that x_0 is consistently replaced with y_0. ∎

Proof of Proposition 23. i) Monotonicity is obvious by definition. Superadditivity holds because —by definition of $E_3(.)$— for any integer $h \geq 3$, h-admissible non-trivial pair (N, X), and strong h-sized weight profile $w \in W^S(h, N)$, $A \in E_3(w, (N, X); h)$, $B \in E_3(w, (N, X); h)$, and $S \cap T = \emptyset$ imply that either $X \in \{A, B\}$ or $x_0 \in A \cap B$. Now, if $X \in \{A, B\}$ then $A \cap B \in \{A, B\}$, hence $A \cap B \in E_3(w, (N, X); h)(S \cup T)$ (by monotonicity). On the other hand, if $X \notin \{A, B\}$ and $x_0 \in A \cap B$ then —by definition of $W^S(h, N)$— it must be the case that $1 \in S \cup T$, whence $A \cap B \in E_3(w, (N, X); h)(S \cup T)$.

To check lack of maximality just take an integer $h \geq 3$, a h-admissible non-trivial pair (N, X), a strong weight profile $w \in W^S(h, N)$, a coalition $S \subseteq N$ such that $1 \notin S$, $w(S) \geq \lfloor h/2 \rfloor + 1$, and notice that —by definition— $\{x_0\} \notin E_3(w, (N, X); h)$ and $B \notin E_3(w, (N, X); h)(N \backslash S)$ for any B such that $x_0 \notin B$.

Instability may be established by the following example of a lower cycle of $E_3(w, (N, X); h)$ with $h = \#N = 5$, $X = \{x_0, y_0, z, v\}$ and $w = (1, 1, 1, 1, 1)$ (see also the proof of Proposition 19 above):

$$\mathbf{C} = ((\{1, 2, 3\}, \{z\}), (\{1, 2, 4\}, \{v\}), (\{3, 4, 5\}, \{x_0, y_0\})).$$

ii) Part ii) follows from the same argument provided in the proof of Proposition 19 ii) .   ∎

Proof of Proposition 25. Again, monotonicity is immediate, by definition. To check for superadditivity, notice that, for any integer $h \geq 3$, h-admissible non-trivial pair (N, X), and weight profile $w \in W^S(h, N)$, $A \in E_4(w, (N, X); h)(S)$, $B \in E_4(w, (N, X); h)(T)$ and $S \cap T = \emptyset$ entail —by definition of $E_4(.)$ and $W^S(.)$—that either $A \cap B \in \{A, B\}$ (under which case $A \cap B \in E_4(w, (N, X); h)$ by monotonicity) or $[x_0 \in A \cap B$, $1 \in (S \cup T) \backslash (S \cap T)$ and either $w(S) \geq \lfloor h/2 \rfloor + 1$ or $w(T) \geq \lfloor h/2 \rfloor + 1]$ (under which case, $A \cap B \in E_4(w, (N, X); h)(S \cup T)$ clearly follows from the definition of $E_4(w, (N, X); h)$).

Concerning maximality, suppose $B \notin E_4(w, (N, X); h)(S)$. If $S = \emptyset$ or $B = \emptyset$ then there is nothing to prove. Thus, suppose $B \neq \emptyset$ and $S \neq \emptyset$ (hence $B \neq X$ and $S \neq N$ as well). Two cases are to be distinguished. If $x_0 \in B$, it must be the case that $1 \notin S$ *and* $w(S) < \lfloor h/2 \rfloor + 1$. Therefore, $X \backslash B \in E_4(w, (N, X); h)(N \backslash S)$ since $1 \in N \backslash S$ and $w(N \backslash S) \geq \lfloor h/2 \rfloor + 1$. If $x_0 \notin B$ then either $1 \notin S$ or $w(S) < \lfloor h/2 \rfloor + 1$: in both cases —by definition of $E_4(.)$— $X \backslash B \in E_4(w, (N, X); h)(N \backslash S)$.

Instability can be immediately established by means of the following example of a lower cycle of $E_4(.)$ (see also the proof of Proposition 19 above): take the sequence $\mathbf{C}'$ obtained from $\mathbf{C}$ as defined in the proof of Proposition 23 above by substituting $\{x_0\}$ for $\{x_0, y_0\}$.   ∎

Proof of Proposition 27. i) Monotonicity follows immediately from the definitions. Superadditivity is readily shown: for any integer $h \geq 3$, h-admissible non-trivial pair (N, X), and strong weight profile $w \in W^S(h, N)$, if $A \in E_5(w, (N, X); h)(S)$, $B \in E_5(w, (N, X); h)(T)$ and $S \cap T = \emptyset$, then —by definition of $E_5(.)$ and of $W^S(.)$— it must be the case that $X \in \{A, B\}$, whence $A \cap B \in \{A, B\}$. Thus, $A \cap B \in E_5(w, (N, X); h)(S \cup T)$ by monotonicity.

Lack of maximality can be established by means of the following class of examples: for any $x \in X$, $S \subseteq N$ such that $1 \notin S$ and $w(S) \geq \lfloor h/2 \rfloor + 1$, both $\{x\} \notin E_5(w, (N, X); h)(S)$ and $B \notin E_5(w, (N, X); h)(N \backslash S)$ for any $B \subseteq X$ such that $x \notin B$.

Stability of $E_5(.)$ follows from its acyclicity or absence of cycles (a property that, as mentioned in the text, is equivalent to core-stability (see e.g. Abdou, Keiding (1991), Kolpin (1991)). To check for acyclicity, let us assume that a cycle $\mathbf{C}^* = ((S_i, B_i))_i$ $i = 1, .., k$ of $E_5(w, (N, X); h)$ with attached sequence (C_i) $i = 1, .., k$. Then, $\mathbf{C}^*$ is such that $1 \in S_i$ for any $i = 1, .., k$ (we can assume w.l.o.g. that the cycle only involves $B_i \neq X$). It follows that—by definition of $\mathbf{C}^*$ and of the sequence (C_i) $i = 1, .., k$ — there is a preference relation $\succ_1$ consistent with $D^0(N, X; h)$ that has no maximal element (indeed, $\succ_1$ is cyclic if X is finite): a contradiction.

Finally, lack of strong stability of $E_5(.)$ is readily checked by adapting the following example due to Demange (1987) (see also Abdou, Keiding (1991)): take $N = \{1, 2, 3, 4\}$, $h = 4$, $w = (1, 1, 1, 1)$, $X = \{x_0, y, z\}$, and the following preference profile $\succ$:

$((z \succ_1 x_0 \succ_1 y), (z \succ_2 x_0 \succ_2 y), (y \succ_3 z \succ_3 x_0), (x_0 \succ_4 y \succ_4 z))$.

Clearly, x_0 is $E_5(w, (N, X); h)$−dominated at profile $\succ$ via $(\{1, 2, 3\}, \{z\})$, and y is $E_5(w, (N, X); h)$−dominated at profile $\succ$ via $(\{1, 2, 4\}, \{x_0\})$, while z is not $E_5(w, (N, X); h)$−dominated at profile $\succ$.

Thus, $C(E_5(w, (N, X); h), \succ) = \{z\}$. However, there is no $S \subseteq N$ such that x_0 is $E_5(w, (N, X); h)$-dominated at profile $\succ$ via $(S, \{z\})$.

ii) See the proof of Proposition 19 ii) .   ■

Proof of Proposition 29. Monotonicity is obvious, and maximality is easily established as follows. Take an integer $h \geq 3$, a h-admissible non-trivial pair (N, X), a strong weight profile $w \in W^S(h, N)$, a fixed majority M for w with $1 \in M$, and $S \subseteq N, B \subseteq X$ such that $B \notin E_6(w, (N, X), M; h)(S)$, and assume —w.l.o.g.—that $S \notin \{N, \emptyset\}$, $B \notin \{X, \emptyset\}$. Then either $x_0 \notin B$ and $(N \backslash S) \cap M \neq \emptyset$ or $x_0 \in B$ and $S \cap M = \emptyset$. In both cases $(X \backslash B) \in E_6(w, (N, X), M; h)(N \backslash S)$.

It remains to be checked that $E_6(.)$ is convex.

Indeed, let $A \in E_6(w, (N, X), M; h)(S)$, $B \in E_6(w, (N, X), M; h)(T)$. Then, one of the following cases holds: i) $x_0 \in A \cap B$, $M \cap S \neq \emptyset$ and $M \subseteq T$; ii)$x_0 \in A \backslash B$, $M \cap S \neq \emptyset$ and $M \subseteq T$; iii) $x_0 \in B \backslash A$, $M \cap T \neq \emptyset$, and $M \subseteq S$; iv) $M \subseteq S$ and $M \subseteq T$.

Under case i): $x_0 \in A \cap B \in E_6(w, (N, X), M; h)(S \cup T)$. If ii) or iii) holds, $M \cap (S \cap T) \neq \emptyset$, hence: $x_0 \in A \cup B \in E_6(w, (N, X), M; h)(S \cap T)$. Finally, under case iv): $M \subseteq S \cap T$: thus $A \cup B \in E_6(w, (N, X), M; h)(S \cap T)$.   ■

Acknowledgement: The paper has greatly benefited from helpful observations and criticisms by two anonymous referees.

References

Abdou J., Keiding H. (1991):*Effectivity Functions in Social Choice.* Dordrecht: Kluwer.

Demange G. (1987): Nonmanipulable Cores. Econometrica 55, 1057-1074.

Kolpin V. (1991): Mixed Effectivity and the Essence of Stability. Social Choice and Welfare 8, 51-63.

Otten G.J., Borm P., Storcken T., Tijs S. (1995): Effectivity Functions and Associated Claim Game Correspondences. Games and Economic Behaviour 9, 172-190.

Peleg B. (1981): Coalition Formation in Simple Games with Dominant Players. International Journal of Game Theory 10, 11-33.

Peleg B. (1984): *Game Theoretic Analysis of Voting in Committees.* Cambridge: Cambridge University Press.

Schofield N. (1985): *Social Choice and Democracy.* Berlin: Springer.

CHAPTER 16

SEQUENTIAL PRODUCTION SITUATIONS AND POTENTIALS

MARK VOORNEVELD (m.voorneveld@kub.nl)
and
STEF TIJS (s.h.tijs@kub.nl)
Department of Econometrics and CentER
Tilburg University
P.O.Box 90153, 5000 LE Tilburg, The Netherlands

LINA MALLOZZI (mallozzi@matna2.dma.unina.it)
Dipartimento di Matematica e Applicazioni R. Caccioppoli
University of Naples Federico II
Naples, Italy

Abstract. This paper studies an important type of production problems in which production takes place in several stages. These problems are modeled as sequential production games, a specific class of extensive form games with imperfect information. These games are related to potential games. Despite the presence of imperfect information, it is shown that pure strategy subgame perfect Nash equilibria do exist, thus allowing for easily adoptable recommendations to firms seeking the advice of game theorists.

1. Introduction

In recent years there has been a growing effort in the study of specific, practically relevant classes of noncooperative games possessing pure strategy Nash equilibria. Rosenthal (1973), for instance, establishes existence of pure Nash equilibria in a congestion model where the players make use of a set of facilities and experience costs due to crowding effects, expressed in terms of a function depending on the number of users of each facility. The paper of Monderer and Shapley (1996), preliminary versions of which had been circulating among game theorists as early as 1988, renewed the interest in this article. They showed that the existence of pure strategy Nash equilibria in the Rosenthal congestion model follows from the existence of a so-called potential function, which summarizes the strategic possibilities of all players in a single real-valued function on the strategy space. Moreover, they showed that each finite potential game is isomorphic to a Rosenthal congestion game.

The work of Monderer and Shapley inspired a large number of papers on congestion effects in games; cf. Borm *et al.* (1997), Holzman and Law-Yone (1997), Konishi *et al.* (1997), Milchtaich (1996), Quint and Shubik (1994). Each focusses on different aspects of a congestion model.

The purpose of the present paper is to describe a production game that often occurs in practice. In this game, raw materials are transformed into a product. The value of the product depends on the activities performed on the raw materials and is divided over the production departments. The production consists of several stages. In each stage, production departments observe the production techniques chosen in the earlier stages and simultaneously perform some activities on the intermediate product (or on the raw materials, if we look at the first stage). The fact that within a stage the departments simultaneously and independently choose a production technique introduces imperfect information into the game. Since the state of the intermediate product strongly depends on the production techniques or activities conducted during the preceding stages, the production departments incur set-up or production costs depending on the previous stages and — of course — on the production strategies of the departments simultaneously performing their activities.

The model is introduced by means of a practical example, based on the processing of rough diamonds. The use of diamond essentially falls into two categories. First of all, properly processed diamond as loose gemstones or part of jewelry has an ornamental function. Secondly, since diamond is the hardest naturally occurring substance, it has an important industrial application: it forms part of cutting and sawing tools, as well as drilling equipment, for instance in mining industry.

In this simplified example, production takes place in two stages and is conducted by three departments. During the first stage, department 1 decides whether a unit of diamond is used for ornamental or industrial purposes, strategies O and I, respectively.

In the second stage, two departments simultaneously perform an activity. In case the unit of diamond was designated for ornamental use, it has to be faceted (cutting flat facets over the entire surface of the stone, usually in a highly symmetrical pattern) and polished to a mirror-like finish to aid light reflection from the surface of the stone or refraction of light through the stone. This is done by department 2, which can use modern equipment to do this (action M), or do the job mostly by relatively old machinery (action O). During the faceting and polishing, department 3 takes care of cooling and lubricating. Department 3 can decide to use high or low quality products to do this, actions *Hi* and *Lo*, respectively.

In case the unit of diamond was designated for industrial use, the second department pulverizes the diamond to produce diamond grit for saw blades. Using the modern action (M) produces grit with a higher mesh (i.e., finer grid, more adequate for precision work) than the old machinery (O). During this process, department 3 takes care of removing debris, again by choosing either high or low quality measures.

The first department operates at negligible costs. In the second stage, departments 2 and 3 clearly incur set-up costs depending on whether processing takes places for ornamental or industrial purposes. These set-up costs are given in Figure 1. Given the industrial or ornamental purpose

Purpose	Set-up dept.2	Set-up dept.3
I	1	1
O	2	3

Figure 1. Set-up costs

decided on in the first stage and the technique (either *Hi* or *Lo*) chosen by the third department, the operating costs of department 2 are given in Figure 2, with a similar specification of the production costs of department 3. Finally, Figure 3 specifies the value of the end product as a function of

Purpose, tech. of dept. 3	Prod. costs of dept. 2	Purpose, tech. of dept. 2	Prod. costs of dept. 3
(I, Hi)	1	(I, M)	2
(I, Lo)	1	(I, O)	1
(O, Hi)	2	(O, M)	1
(O, Lo)	3	(O, O)	3

Figure 2. Production costs

the production techniques. Assuming that the value of the end product is divided equally over the three production departments and subtracting the costs, one obtains the game in Figure 4. Recall that departments two and three are both in the same stage and hence only observe the action taken in stage 1: histories (I, M) and (I, O) are in one information set, just like histories (O, M) and (O, O).

For instance, if the production profile is (I, M, Hi), each department receives one third of 27. Department 1 incurs no costs, so the payoff to this

Production profile	Value
(I, M, Hi)	27
(I, M, Lo)	15
(I, O, Hi)	27
(I, O, Lo)	12
(O, M, Hi)	21
(O, M, Lo)	18
(O, O, Hi)	30
(O, O, Lo)	18

Figure 3. Value of end product

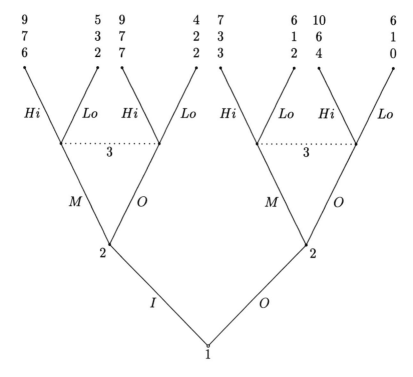

Figure 4. The diamond game

department equals 9. Department 2 incurs set-up costs 1 and production costs 1, so the payoff to this department equals $9 - 1 - 1 = 7$. Similarly, department 3 has payoff $9 - 1 - 2 = 6$ since its set-up costs are 1 and its processing costs are 2.

Sequential production situations give rise to a special class of imperfect information games, since players at a certain stage observe the production techniques chosen in previous stages, but not those of the departments in the same and later stages. Thus, in general, the existence of pure strategy Nash equilibria is not guaranteed. In the diamond game of Figure 4, however, there are several. We show that this is no coincidence and that these games are closely related to potential games. A more formal description of the model is provided in Section 2. Section 3 contains most results. Section 4 concludes with remarks concerning extensions of the model.

2. Model

This section contains a formal description of the model. The games arising from sequential production situations are hierarchical games.

DEFINITION 2.1.   A *hierarchical game* is an extensive form game described by a tuple $H = \langle N = N_1 \cup \cdots \cup N_m, (A_i)_{i \in N}, (u_i)_{i \in N} \rangle$. The finite player set N is an ordered partition $N_1 \cup \cdots \cup N_m$. The number $m \in \mathbf{N}$ denotes the number of stages of the game. For $k \in \{1, \ldots, m\}$, N_k denotes the set of players operating at stage k. Each player $i \in N$ has a finite set A_i of actions containing at least two elements and a payoff function $u_i : \prod_{j \in N} A_j \to \mathbf{R}$. The game is played in such a way that for each stage $k \in \{1, \ldots, m\}$, the players in N_k observe only the action choices of the players in $N_1 \cup \cdots \cup N_{k-1}$ operating in the previous stages and then simultaneously and independently choose an action.

Notice that a hierarchical game is a specific type of extensive form game with imperfect information. The players in N_1, operating in the first stage, make no observations prior to simultaneously and independently choosing their action. The players in N_2, operating in the second stage, observe the actions of the players in N_1 and then simultaneously and independently choose their actions, thus having no information about the action choices of the other players in the same stage and the players in later stages. The same reasoning applies to later stages of the game. Strategic games are a special case, since they can be modelled by a hierarchical game with only one stage: all players simultaneously and independently make a strategy choice.

The players are assumed to be numbered from 1 through $|N|$; players with a low number play in early stages, i.e., if $i \in N_k, j \in N_l$, and $i < j$, then $k \leq l$.

The following notation is used. The *predecessors* $Pr(i)$ of a player $i \in N$ are those players operating at an earlier stage than i:

$$\forall k \in \{1,\dots,m\}, \forall i \in N_k : Pr(i) := \cup_{l \in \{1,\dots,m\}, l<k} N_l.$$

The *colleagues* $C(i)$ of a player $i \in N$ are those players operating at the same stage as player i:

$$\forall k \in \{1,\dots,m\}, \forall i \in N_k : C(i) := N_k \setminus \{i\}.$$

The *followers* $F(i)$ of a player $i \in N$ are those players operating at a later stage than player i:

$$\forall k \in \{1,\dots,m\}, \forall i \in N_k : F(i) := \cup_{l \in \{1,\dots,m\}, l>k} N_l.$$

For instance, in the diamond game of Figure 4, department 1 has no predecessors, no colleagues, and followers 2 and 3. Department 2 has predecessor 1, colleague 3, and no followers.

DEFINITION 2.2.   A *sequential production situation* is a tuple

$$\langle N = N_1 \cup \cdots \cup N_m, (A_i)_{i \in N}, \rho, (c_i)_{i \in N} \rangle,$$

where the set N of production departments or players is described by an ordered partition $N_1 \cup \cdots \cup N_m$. The number $m \in \mathbf{N}$ denotes the number of production stages. Each player $i \in N$ has a finite set A_i of production techniques (containing at least two elements). The function $\rho : \prod_{i \in N} A_i \to \mathbf{R}$ specifies for each production profile $a = (a_i)_{i \in N} \in \prod_{i \in N} A_i$ the value $\rho(a)$ of the end product. Each player $i \in N$ has a cost function c_i denoting the set-up and operating costs of this player. This cost function depends on the predecessors (set-up) and colleagues (operating) (if any), i.e.,

$$\forall i \in N \ : \ c_i : \prod_{j \in Pr(i) \cup C(i)} A_j \to \mathbf{R}.$$

Production takes place in such a way that for each stage $k \in \{1,\dots,m\}$, the players in N_k observe only the production techniques of the players in $N_1 \cup \cdots \cup N_{k-1}$ operating in the previous stages and then simultaneously and independently choose a production technique.

REMARK 2.3.   The definition of c_i for players $i \in N_1$ in the first stage deserves some special attention. In this case, the set of predecessors $Pr(i)$ of i is empty by definition, so c_i is a function only of i's colleagues. If this set also happens to be empty, i.e., if there is only one department i in the first stage, we allow $c_i \in \mathbf{R}$ to be an arbitrary constant.

REMARK 2.4.   Two main assumptions underlie the production process captured by a sequential production situation. In this remark, some motivation for these assumptions is provided.

- The first assumption is that the departments within a production stage independently and simultaneously choose a production technique. Many modern firms are decentralized; departments act as autonomous units with their own decision power. In such environments this assumption is reasonable.
- The second assumption is that the operating costs of a production department do not depend on its own technique. This is equivalent with stating that a production department has fixed operating costs given the state of the intermediate product and the production techniques of the colleagues.

Given a sequential production situation and assuming for now that the value of the end product is split equally over the departments or players (for a relaxation of this assumption, see Section 4), one can easily define its associated hierarchical game.

DEFINITION 2.5.   Let $\langle N = N_1 \cup \cdots \cup N_m, (A_i)_{i \in N}, \rho, (c_i)_{i \in N} \rangle$ be a sequential production situation. The associated *sequential production game* is the hierarchical game $\langle N = N_1 \cup \cdots \cup N_m, (A_i)_{i \in N}, (u_i)_{i \in N} \rangle$ with for all $i \in N$ and all $a \in \prod_{i \in N} A_i$:

$$u_i(a) = \begin{cases} \frac{1}{|N|}\rho(a) - c_i & \text{if } i \in N_1 \text{ and } C(i) = \emptyset \\ \frac{1}{|N|}\rho(a) - c_i((a_j)_{j \in Pr(i) \cup C(i)}) & \text{otherwise} \end{cases}$$

That is, the payoff to a production department is an equal share of the value of the end product minus the costs it incurs.

The strategic game corresponding to a hierachical game $H = \langle N = N_1 \cup \cdots \cup N_m, (A_i)_{i \in N}, (u_i)_{i \in N} \rangle$ is defined — in the usual way — to be the strategic game $\mathcal{N}(H) = \langle N, (S_i)_{i \in N}, (U_i)_{i \in N} \rangle$, called the *normalization* of H, where the strategy space S_i of player $i \in N$ prescribes an action choice in every contingency that a player may be called upon to act and the payoff function associates to each strategy profile the payoff in the outcome of the hierarchical game induced by this strategy. Formally,

$$S_i = \begin{cases} A_i & \text{if } i \in N_1, \\ \{\sigma_i \mid \sigma_i : \prod_{j \in Pr(i)} A_j \to A_i\} & \text{if } i \in N_k, k \geq 2. \end{cases}$$

Inductively, one can define the realized play of the game by means of a function $r : \prod_{i \in N} S_i \to \prod_{i \in N} A_i$ as follows.

$$r_i(\sigma) = \begin{cases} \sigma_i \in A_i & \text{if } i \in N_1, \\ \sigma_i((r_j(\sigma))_{j \in Pr(i)}) & \text{if } i \in N_k, k \geq 2. \end{cases}$$

Player i's payoff function U_i assigns to every strategy profile $\sigma = (\sigma_i)_{i \in N} \in \prod_{i \in N} S_i$ the payoff associated with the outcome realized by σ: $U_i(\sigma) = u_i(r(\sigma))$.

For instance, in the strategic game corresponding to the diamond game of Figure 4, player 1 has two strategies: $S_1 = A_1 = \{I, O\}$. Player 2 has 4 strategies: $S_2 = \{(M, M), (M, O), (O, O), (O, M)\}$, where the first coordinate specifies the action choice if player 1 chose I and the second coordinate specifies the action choice if player 1 chose O. Similarly, the strategy space of player 3 equals $\{(Hi, Hi), (Hi, Lo), (Lo, Lo), (Lo, Hi)\}$. The strategic game is given in Figure 5.

	(Hi, Hi)	(Hi, Lo)	(Lo, Hi)	(Lo, Lo)
(M, M)	9,7,6	9,7,6	5,3,2	5,3,2
(M, O)	9,7,6	9,7,6	5,3,2	5,3,2
(O, M)	9,7,7	9,7,7	4,2,2	4,2,2
(O, O)	9,7,7	9,7,7	4,2,2	4,2,2

Department 1 plays I

	(Hi, Hi)	(Hi, Lo)	(Lo, Hi)	(Lo, Lo)
(M, M)	7,3,3	6,1,2	7,3,3	6,1,2
(M, O)	10,6,4	6,1,0	10,6,4	6,1,0
(O, M)	7,3,3	6,1,2	7,3,3	6,1,2
(O, O)	10,6,4	6,1,0	10,6,4	6,1,0

Department 1 plays O

Figure 5.   The normalization of the diamond game

Some matters of notation. In the normalization of a hierarchical game, the strategy "always choose a_i" is denoted $\bar{a}_i$. Furthermore, conventional game theoretic notation is used. For instance, $S := \prod_{j \in N} S_j$ denotes the set of strategy profiles for all players in N, $S_{-i} := \prod_{j \in N \setminus \{i\}} S_j$ denotes the set of strategy profiles of i's opponents. Similar notation is adopted for elements of these sets: $\sigma \in S, \sigma_{-i} \in S_{-i}$, and for profiles of actions, rather than strategies.

3. Results

In this section the sequential production games are related to exact potential games introduced by Monderer and Shapley (1996). Hierarchical potential games are defined and, analogous to the isomorphism between congestion games à la Rosenthal (1973) and exact potential games, it is shown that not only every sequential production game is a hierarchical potential game, but conversely, every hierarchical potential game can be seen as a well-chosen sequential production game. This result has an important implication: sequential production games have pure strategy equilibria. So-called potential maximizing strategies, introduced in Monderer and Shapley (1996) and studied in more detail by Peleg, Potters, and Tijs (1996), form an interesting equilibrium refinement and are studied for this class of games.

First, recall the definition of exact potential games:

DEFINITION 3.1.   A strategic game $G = \langle N, (S_i)_{i \in N}, (U_i)_{i \in N} \rangle$ is an *exact potential game* if there exists a function $P : \prod_{i \in N} S_i \to \mathbf{R}$ such that for each player $i \in N$, each profile $\sigma_{-i} \in \prod_{j \in N \setminus \{i\}} S_j$ of strategies of the opponents, and each pair of strategies $\sigma_i, \tau_i \in S_i$ of player i:

$$U_i(\sigma_i, \sigma_{-i}) - U_i(\tau_i, \sigma_{-i}) = P(\sigma_i, \sigma_{-i}) - P(\tau_i, \sigma_{-i}),$$

i.e., if the change in the payoff to a unilaterally deviating player is equal to the change in the value of the function P. P is called an *(exact) potential* of the game.

It is easy to see that the set of Nash equilibria of the game G coincides with the set of Nash equilibria of the game $\langle N, (S_i)_{i \in N}, (P)_{i \in N} \rangle$ with all payoff functions replaced by the potential function P. Finite exact potential games consequently have pure strategy Nash equilibria: the potential P achieves a maximum over the finite set $\prod_{i \in N} S_i$, which is easily seen to be a pure strategy Nash equilibrium.

Facchini *et al.* (1997, Theorem 2.1) showed that a game is an exact potential game if and only if there exists a real-valued function P on the strategy space such that for each player i, the difference between his payoff and the function P does not depend on the strategy choice of player i himself. That is, an exact potential game can be seen as the 'sum' of a coordination game, in which the payoff to all players is given by the function P, and a dummy game, in which the payoff to a player is independent of his own strategy choice. This result is used later, so we summarize it in a lemma.

LEMMA 3.2.   *A strategic game* $G = \langle N, (S_i)_{i \in N}, (U_i)_{i \in N} \rangle$ *is an exact potential game if and only if there exists a function* $P : \prod_{i \in N} S_i \to \mathbf{R}$ *and for each player* $i \in N$ *a function* $D_i : \prod_{j \in N \setminus \{i\}} S_j \to \mathbf{R}$ *such that*

$$\forall i \in N, \forall \sigma \in \prod_{j \in N} S_j : U_i(\sigma) - P(\sigma) = D_i(\sigma_{-i}).$$

The function P in Lemma 3.2 is easily seen to be an exact potential of the game.

If the normalization of a hierarchical game is a potential game, then the potential depends on the realized outcome, but not on the strategies leading to this outcome:

LEMMA 3.3. *Let H be a hierarchical game. If its normalization $\mathcal{N}(H)$ is a potential game with potential function P, and σ, τ are strategy profiles such that $r(\sigma) = r(\tau)$, then $P(\sigma) = P(\tau)$.*

Proof. If $r(\sigma) = r(\tau) = (a_j)_{j \in N}$, then $\sigma_i = \tau_i$ for all $i \in N_1$ and for players $i \in N_k, k \geq 2$, σ_i and τ_i differ only in their behavior off the play path $(a_j)_{j \in N}$. Thus, the payoff in $\mathcal{N}(H)$ to deviating players along the path from $\sigma = (\sigma_1, \ldots, \sigma_n)$ to $(\tau_1, \sigma_2, \ldots, \sigma_n)$ to ... to $(\tau_1, \ldots, \tau_{n-1}, \sigma_n)$ to $\tau = (\tau_1, \ldots, \tau_n)$ does not change. Hence $P(\sigma) = P(\tau)$. ∎

DEFINITION 3.4. A hierarchical game $H = \langle N = N_1 \cup \cdots \cup N_m, (A_i)_{i \in N}, (u_i)_{i \in N}$ is called a *hierarchical potential game* if there exist functions $p : \prod_{j \in N} A_j \to \mathbf{R}$ and $(d_i)_{i \in N}$ with

$$\forall i \in N : d_i : \prod_{j \in Pr(i) \cup C(i)} A_j \to \mathbf{R},$$

or $d_i \in \mathbf{R}$ if $i \in N_1$ and $C(i) = \emptyset^1$, such that for each player $i \in N$ and each action profile $a \in \prod_{i \in N} A_i$:

$$u_i(a) = \begin{cases} p(a) + d_i & \text{if } i \in N_1 \text{ and } C(i) = \emptyset, \\ p(a) + d_i((a_j)_{j \in Pr(i) \cup C(i)}) & \text{otherwise.} \end{cases}$$

The function p is called a *potential* for H.

The reason for this definition is the following:

THEOREM 3.5. *A hierarchical game $H = \langle N = N_1 \cup \cdots \cup N_m, (A_i)_{i \in N}, (u_i)_{i \in N} \rangle$ is a hierarchical potential game if and only if its normalization $\mathcal{N}(H) = \langle N, (S_i)_{i \in N}, (U_i)_{i \in N} \rangle$ is an exact potential game.*

Proof. If H is a hierarchical potential game with $p, (d_i)_{i \in N}$ as in Definition 3.4, then by definition of the normalized game one has that for each $\sigma \in \prod_{i \in N} S_i$:

$$U_i(\sigma) = u_i(r(\sigma)) = p(r(\sigma)) + d_i((r_j(\sigma))_{j \in Pr(i) \cup C(i)}).$$

Lemma 3.2 implies that $\mathcal{N}(H)$ is an exact potential game.

To prove the converse, assume $\mathcal{N}(H)$ is an exact potential game with potential P. We have to show the existence of functions p and $(d_i)_{i \in N}$ as in Definition 3.4. For each $a \in \prod_{i \in N} A_i$, recall that $\bar{a}_i \in S_i$ is the strategy in

---

[1] Analogous to Remark 2.3.

which player i always chooses a_i. Denote $\bar{a} = (\bar{a}_i)_{i \in N}$. Define $p(a) = P(\bar{a})$. The definition of the $(d_i)_{i \in N}$ is split up into two cases.

CASE 1: $i \in N_m$. Lemma 3.2 implies the existence of a function $D_i :$ $\prod_{j \in N \setminus \{i\}} S_j \to \mathbf{R}$ such that $U_i(\sigma) = P(\sigma) + D_i(\sigma_{-i})$ for each $\sigma \in \prod_{j \in N} S_j$. Define for each $a_{-i} \in A_{-i}$: $d_i(a_{-i}) = D_i(\bar{a}_{-i})$. Then, for each $a \in \prod_{j \in N} A_j$, $u_i(a) = U_i(\bar{a}) = P(\bar{a}) + D_i(\bar{a}_{-i}) = p(a) + d_i(a_{-i}) = p(a) + d_i((a_j)_{j \in Pr(i) \cup C(i)})$.

CASE 2: $i \in N_k, k < m$. To prove the existence of d_i as in Definition 3.4, it suffices to show that $u_i - p$ does not depend on the actions chosen by player i himself and i's followers, since we can then take d_i equal to this difference. Formally, it is shown that for all $a \in A$ and $(b_j)_{j \in F(i) \cup \{i\}} \in \prod_{j \in F(i) \cup \{i\}} A_j$:

$$
\begin{aligned}
u_i(a) - p(a) &= u_i((a_j)_{j \in Pr(i) \cup C(i)}, (b_j)_{j \in F(i) \cup \{i\}}) \\
&\quad - p((a_j)_{j \in Pr(i) \cup C(i)}, (b_j)_{j \in F(i) \cup \{i\}}). \tag{1}
\end{aligned}
$$

Let $a \in A$ and $(b_j)_{j \in F(i) \cup \{i\}} \in \prod_{j \in F(i) \cup \{i\}} A_j$.

CASE 2A: Suppose $a_i \neq b_i$. Define

- $\sigma_i = \bar{a}_i$ and $\tau_i = \bar{b}_i$,
- for each player $j \in Pr(i) \cup C(i)$: $\sigma_j = \bar{a}_j$,
- for $j \in F(i)$, let σ_j be the strategy that always chooses b_j, unless the history is $(a_k)_{k \in Pr(j)}$, in which case j chooses a_j.

Notice that $r(\sigma_i, \sigma_{-i}) = a, r(\tau_i, \sigma_{-i}) = ((a_j)_{j \in Pr(i) \cup C(i)}, (b_j)_{j \in F(i) \cup \{i\}})$. By Lemma 3.2: $U_i(\sigma) - P(\sigma) = U_i(\tau_i, \sigma_{-i}) - P(\tau_i, \sigma_{-i})$. By Lemma 3.3: $P(\eta) = P(\tau)$ if $r(\eta) = r(\tau)$. Hence

$$
\begin{aligned}
u_i(a) - p(a) &= u_i(r(\sigma_i, \sigma_{-i})) - p(r(\sigma_i, \sigma_{-i})) \\
&= U_i(\sigma_i, \sigma_{-i}) - P(\bar{a}) \\
&= U_i(\sigma_i, \sigma_{-i}) - P(\sigma_i, \sigma_{-i}) \\
&= U_i(\tau_i, \sigma_{-i}) - P(\tau_i, \sigma_{-i}) \\
&= U_i(\tau_i, \sigma_{-i}) - P((\bar{a}_j)_{j \in Pr(i) \cup C(i)}, (\bar{b}_j)_{j \in F(i) \cup \{i\}}) \\
&= u_i(r(\tau_i, \sigma_{-i})) - p((a_j)_{j \in Pr(i) \cup C(i)}, (b_j)_{j \in F(i) \cup \{i\}}) \\
&= u_i((a_j)_{j \in Pr(i) \cup C(i)}, (b_j)_{j \in F(i) \cup \{i\}}) \\
&\quad - p((a_j)_{j \in Pr(i) \cup C(i)}, (b_j)_{j \in F(i) \cup \{i\}})
\end{aligned}
$$

which proves that (1) holds if $b_i \neq a_i$.

CASE 2B: Suppose $a_i = b_i$. By assumption (cf. Definition 2.1), A_i contains at least two elements. Let $c_i \in A_i$ with $c_i \neq a_i$. Applying the result of case

2A twice yields:

$$
\begin{aligned}
u_i(a) - p(a) &= u_i((a_j)_{j \in Pr(i) \cup C(i)}, c_i, (b_j)_{j \in F(i)}) \\
&\quad - p((a_j)_{j \in Pr(i) \cup C(i)}, c_i, (b_j)_{j \in F(i)}) \\
&= u_i((a_j)_{j \in Pr(i) \cup C(i)}, b_i, (b_j)_{j \in F(i)}) \\
&\quad - p((a_j)_{j \in Pr(i) \cup C(i)}, b_i, (b_j)_{j \in F(i)}) \\
&= u_i((a_j)_{j \in Pr(i) \cup C(i)}, (b_j)_{j \in F(i) \cup \{i\}}) \\
&\quad - p((a_j)_{j \in Pr(i) \cup C(i)}, (b_j)_{j \in F(i) \cup \{i\}}),
\end{aligned}
$$

which proves that (1) holds if $b_i = a_i$.   ∎

The assumption in Definition 2.1 that each player $i \in N$ has an action set A_i containing at least two elements is relatively innocent: players having to make a choice from a singleton set of options are not extremely interesting. Notice, however, that in the proof above we explicitly made use of this assumption. In fact, the following example shows that the 'if'-part of Theorem 3.5 *breaks down* if some of the players have only one action.

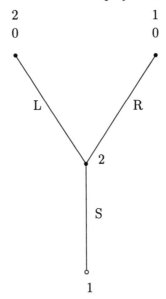

Figure 6.   Player 1 has only one action

EXAMPLE 3.6.   Consider the extensive form game in Figure 6 where player 1 has only one action S and player 2 in the next stage chooses either L or R. Payoffs are $u_1(S, L) = 2, u_1(S, R) = 1, u_2(S, L) = u_2(S, R) = 0$. Its normalization is clearly an exact potential game. But $p : \{S\} \times \{L, R\} \rightarrow$

$\mathbf{R}, d_1 \in \mathbf{R}$, and $d_2 : \{S\} \to \mathbf{R}$ as in Definition 3.4 would have to satisfy the following inconsistent system of linear equations:

$$\begin{cases} u_1(S, L) = p(S, L) + d_1 \\ u_1(S, R) = p(S, R) + d_1 \\ u_2(S, L) = p(S, L) + d_2(S) \\ u_2(S, R) = p(S, R) + d_2(S) \end{cases}$$

The last two equations imply that p has to be a constant function. But then $u_1 - p$ depends on the action choice of player 2. In hierarchical potential games, the difference between the payoff function and a potential was assumed to be independent of the action choices of followers.

The following theorem relates hierarchical games and sequential production games.

THEOREM 3.7. *Every sequential production game is a hierarchical potential game. For every hierarchical potential game there is a sequential production situation that induces this game.*

Proof. To see that a sequential production game as in Definition 2.5 is a hierarchical potential game, take

$$p = \tfrac{1}{|N|} \rho,$$
$$d_i = -c_i \quad \text{for each } i \in N.$$

To prove the converse, consider a hierarchical potential game $\langle N = N_1 \cup \cdots \cup N_m, (A_i)_{i \in N}, (u_i)_{i \in N} \rangle$ with p and $(d_i)_{i \in N}$ as in Definition 3.4. Then the sequential production situation $\langle N = N_1 \cup \cdots N_m, (A_i)_{i \in N}, \rho, (c_i)_{i \in N} \rangle$ with

$$\rho = |N| p,$$
$$c_i = -d_i \quad \text{for each } i \in N$$

induces exactly the same game. ∎

Notice that a potential of a sequential production game equals the value function ρ divided by the number of players.

After defining hierarchical games we observed that every finite strategic game can be seen an a hierarchical game with only one stage. The theorem above establishes that every exact potential game is essentially a hierarchical potential game or a sequential production game.

It follows from the remark after the definition of potential games that every hierarchical potential game and thus every sequential production game has a pure strategy Nash equilibrium. One can even extend this result to subgame perfect equilibria, as is done below.

Subgames of imperfect information games are defined as usual. In hierarchical games, this implies that the game itself is a subgame, and that for each number k of stages, each profile of actions of the players in the first k stages induces a subgame. Formally,

DEFINITION 3.8.   Let $H = \langle N = N_1 \cup \cdots \cup N_m, (A_i)_{i \in N}, (u_i)_{i \in N} \rangle$ be a hierarchical game. Then H itself is a subgame and, moreover, for each $k \in \{1, \ldots, m-1\}$ and each profile or history $h = (a_i)_{i \in N_1 \cup \ldots \cup N_k} \in \prod_{i \in N_1 \cup \ldots \cup N_k} A_i$, the subgame $H(h)$ is the hierarchical game $\langle N_{k+1} \cup \cdots \cup N_m, (A_i)_{i \in N_{k+1} \cup \cdots \cup N_m}, (\tilde{u}_i)_{i \in N_{k+1} \cup \cdots \cup N_m} \rangle$ with

$$\tilde{u}_i(\cdot) = u_i(h, \cdot) \text{ for each } i \in N_{k+1} \cup \cdots \cup N_m.$$

For instance, the subgame $H(I)$ that arises if department 1 chooses action I in the diamond game is given in Figure 7. Realize that since the remaining departments both operate at the same stage, the subgame $H(I)$ has no subgame other than itself.

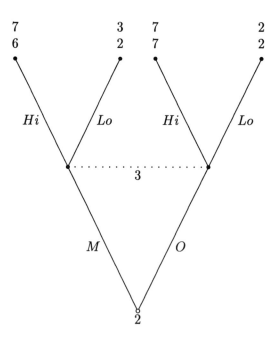

Figure 7.   The subgame $H(I)$ of the diamond game

COROLLARY 3.9.   *Every subgame of a hierarchical potential game is a hierarchical potential game.*

This corollary can be proven either directly, using Definition 3.4, or indirectly, using Theorem 3.5. The details are left to the reader. Notice that if p is a potential of the hierarchical game H, then for each subgame $H(h)$, the function $\tilde{p}$ with $\tilde{p}(\cdot) = p(h, \cdot)$ is a potential for the subgame $H(h)$.

Recall that a strategy profile σ in the normalized game $\mathcal{N}(H)$ is a *subgame perfect Nash equilibrium* if it induces a Nash equilibrium in each subgame, i.e., if behavior outside the play path is also credible. For instance, $(O, (O, O), (Hi, Hi))$ is a subgame perfect Nash equilibrium of the diamond game, but $(O, (M, O), (Lo, Hi))$ is not, since in the subgame $H(I)$ of Figure 7 player 3 would rather play Hi than Lo.

Potential maximizing strategies form a refinement of the Nash equilibrium concept in strategic games with a potential. This refinement was introduced by Monderer and Shapley (1996). It was studied axiomatically in Peleg, Potters, and Tijs (1996) and used in Voorneveld (1997) to derive equilibrium existence results in infinite games. In hierarchical potential games H the notion of potential maximizing strategies can be extended to *subgame potential maximizing strategies*, being those strategy profiles σ in the normalization $\mathcal{N}(H)$ that select actions maximizing the potential in every subgame. Corollary 3.9 guarantees that subgame potential maximizers are well-defined.

The next theorem establishes one of the main results of this paper: hierarchical potential games, and in particular sequential production games, have subgame perfect Nash equilibria in pure strategies, despite the presence of imperfect information.

THEOREM 3.10.   *Let H be a hierarchical potential game and $\mathcal{N}(H)$ its normalization.*

- $\mathcal{N}(H)$ *has a subgame potential maximizing strategy profile in pure strategies;*
- *each such pure strategy subgame potential maximizing profile is a pure strategy subgame perfect Nash equilibrium;*
- *not every pure strategy subgame perfect Nash equilibrium is a pure strategy subgame potential maximizer.*

Proof. The proof of the first claim proceeds by induction on the number of stages of the game and closely mimics the existence proof of pure strategy Nash equilibria in standard perfect information games. It is therefore left to the reader.

Strategies maximizing the potential of a subgame are easily seen to be Nash equilibrium strategies for the subgame by using Definition 3.4: the only difference between u_i and the potential p is a function d_i not depending on the choices of player i. This proves the second claim.

The final claim already follows from the insights in potential games in strategic form. Consider the single stage hierarchical potential game H with player set $N = N_1 = \{1, 2\}$, action sets $A_1 = A_2 = \{\alpha, \beta\}$, potential $p : A_1 \times A_2 \to \mathbf{R}$ with $p(\alpha, \alpha) = 2, p(\alpha, \beta) = p(\beta, \alpha) = 0, p(\beta, \beta) = 1$ and with $d_1 : A_2 \to \mathbf{R}$ and $d_2 : A_1 \to \mathbf{R}$ equal to the zero function. This is just the 2×2 exact potential game in Figure 8. Notice that (β, β) is a pure

	α	β
α	2,2	0,0
β	0,0	1,1

Figure 8. (β, β) subgame perfect, not potential maximizing.

strategy subgame perfect Nash equilibrium (there is only one subgame, namely the game itself), but not potential maximizing.   ∎

In the diamond game of Figure 4, the pure strategy subgame potential maximizers are $(O, (M, O), (Hi, Hi))$ and $(O, (O, O), (Hi, Hi))$. The profile $(O, (M, O), (Lo, Hi))$ is potential maximizing, but does not select a potential maximizing outcome in the subgame $H(I)$.

Peleg, Potters, and Tijs (1996), and Peleg and Tijs (1996) give axiomatizations of several solution concepts in strategic games. By straightforward extension of their theorems, it is possible to axiomatize the subgame potential maximizing strategies introduced in this paper. We decided not to include this result; it seems too technical for a paper on game practice, it requires considerable space to properly introduce the axioms, and the proof closely resembles those of the previously mentioned authors.

4. Conclusions and extensions of the model

Practical situations can sometimes be studied using game theoretic tools. The topic of this paper has been the study of an important type of production problems in which production takes place in several stages. These problems were modeled as sequential production games, a specific class of extensive form games with imperfect information. These games were related to potential games. In fact, it was shown that the class of sequential production games coincides with the class of hierarchical potential games (cf. Theorem 3.7).

Firms seeking the help from game theorists want clear-cut recommendations. Extensive form games with incomplete information typically do not have pure strategy equilibria, which makes it hard to provide such easily adoptable recommendations. A significant feature of sequential production games is the existence of pure strategy subgame perfect Nash equilibria. Using subgame potential maximizing profiles, we were able to identify a subset of these equilibria. It is worth mentioning that subgame potential maximizing strategies can be computed as soon as the potential p of a hierarchical potential game H or the value function ρ in a sequential production situation is known.

In Definition 2.5, payoffs to departments in a sequential production game were determined by giving each department an *equal* share of the value of the end product and then subtracting the costs. A possible extension of the model is to consider unequal division of the value over the departments. Introduce a vector $(w_i)_{i \in N}$ of weights satisfying $w_i \geq 0$ for each department i and such that $\sum_{i \in N} w_i = 1$. The payoff functions u_i in Definition 2.5 can then be changed to $u_i = w_i \rho - c_i$.

Such unequal splitting of the value of the end product might be reasonable in the following sequential production situation. Students of a graduate school, the 'raw materials', receive an education in three 'production stages': there are preliminary or refresher courses in the first stage, the core courses in the second stage, and specialized courses in the third stage. The value of the 'end product', the PhD student successfully finishing the three stages, is usually considered to be the result of the specialized, advanced courses, to a lesser degree of the core courses, and hardly of the preliminary and refresher courses. In this teaching example, it appears reasonable to measure the contribution to the end product in such a way that a larger weight is assigned to lecturers teaching more advanced material.

Making the necessary modifications, the main results of this paper still hold for sequential production games with unequal splitting of the value over the production departments. In particular, pure strategy subgame perfect Nash equilibria still exist.

The class of games generated in this way is closely related to weighted potential games, a class of ordinal potential games introduced in Monderer and Shapley (1996). Ordinal potential games were characterized in Voorneveld and Norde (1997). For another practical class of ordinal potential games, refer to Voorneveld, Koster, and Reijnierse (1998), who consider schemes to finance public goods in a voluntary contribution game.

Sequential production situations have an ordered partition of the player set, specifying which departments or players acted in which stage. Another modification is to relax this assumption and to have departments which act at most once in every possible play path. If in each path the costs of a department depend only on colleagues — players acting in the same information set — and predecessors, similar results arise: such games can be related to exact potential games and the existence of pure strategy subgame perfect Nash equilibria can be established.

References

Borm P., Facchini G., Van Megen F., Tijs S., and Voorneveld M. (1997) Strong Nash equilibria and the potential maximizer, mimeo, Tilburg University.

Facchini G., Van Megen F., Borm P., and Tijs S. (1997) Congestion models and weighted Bayesian potential games, *Theory and Decision*, **42**, pp. 193–206.

Holzman R. and Law-Yone N. (1997) Strong equilibrium in congestion games, *Games and Economic Behavior*, **21**, pp. 85–101.

Konishi H., Le Breton M., and Weber S. (1997) Equilibrium in a model with partial rivalry, *Journal of Economic Theory*, **72**, pp. 225–237.

Milchtaich I. (1996) Congestion models with player specific payoff functions, *Games and Economic Behavior*, **13**, pp. 111–124.

Monderer D. and Shapley L.S. (1996) Potential games, *Games and Economic Behavior*, **14**, pp. 124–143.

Peleg B., Potters J., and Tijs S. (1996) Minimality of consistent solutions for strategic games, in particular for potential games, *Economic Theory*, **7**, pp. 81–93.

Peleg B. and Tijs S. (1996) The consistency principle for games in strategic form, *International Journal of Game Theory*, **25**, pp. 13–34.

Quint T. and Shubik M. (1994) A model of migration, working paper, Cowles Foundation, Yale University.

Rosenthal R.W. (1973) A class of games possessing pure strategy Nash equilibria, *International Journal of Game Theory*, **2**, pp. 65–67.

Voorneveld M., Koster M., and Reijnierse H. (1998) Voluntary contribution to multiple facilities; a class of ordinal potential games, mimeo, Tilburg University.

Voorneveld M. and Norde H. (1997) A characterization of ordinal potential games, *Games and Economic Behavior*, **19**, pp. 235–242.

Voorneveld M. (1997) Equilibria and approximate equilibria in infinite potential games, *Economics Letters*, **56**, pp. 163–169.

CHAPTER 17

APPROXIMATE ENVY-FREE PROCEDURES

DAO-ZHI ZENG (zeng@ec.kagawa-u.ac.jp)
Faculty of Economics, Kagawa University
Takamatsu, Kagawa, 760-8523 Japan

Abstract. Consider a situation in which n players divide a cake. Each player has a preference expressed by a measure on the cake. Given a positive number ϵ, a cake division is said to be ϵ-envy-free, if every player measures his/her piece not smaller than the largest piece by ϵ. This paper first forms an ϵ-envy-free procedure for such an approximate envy-free division, based on a known idea of Brams and Taylor (1996), then presents another completely new procedure. The first procedure does not work for chore division but the second one works well. The number of necessary cuts of each procedure is bounded. Furthermore, the new procedure is generalized for ϵ-multi-fair division. Finally, this paper gives procedures for ϵ-envy-free division in unequal ratios.

1. Introduction

Steinhaus (1948) raised the problem of dividing a cake fairly among n players, which attracts many mathematicians, economists and political scientists because of its wide application. There are many criteria of fairness. This paper focus on a particular criterion of fairness, namely, envy-freeness. A presupposition is that players may have different preferences, expressed by measures on the cake. A cake division is said to be envy-free if each player measures his/her assigned piece largest.

In contrast to dividing a delicious cake, sometimes people divide chores or dirty work. In such a situation, every player prefers a smaller piece. We call this kind of problems as chore divisions. Many, but not all, results for cake division hold similarly for chore division (see Peterson and Su (1988)).

We are interested in designing a procedure for fair division, which consists of rules and strategies for players. An envy-free procedure outputs an envy-free division if n players divide the whole cake (or the set of chores) according to the rules step by step, using the strategies suggested in the procedure. Brams and Taylor (1995) give a finite discrete procedure (for the definitions of various procedures, see Brams and Taylor (1996)) for envy-free divisions. A shortcoming of their procedure is that, although the number of cuts necessary for obtaining an envy-free division by that procedure is finite, it could be arbitrarily large by a suitable choice of the measures corresponding to the players' preferences.

Brams, Taylor and Zwicker (1997) provide a moving-knife procedure for four players' envy-free division. The number of cuts is bounded by eleven. However, the procedure has not been generalized to a division among more than four players, if we do not use any "oracle". It is still an open question whether there is a procedure for envy-free divisions among n-players with bounded number of cuts.

This paper considers an ϵ-envy-free division for any $\epsilon > 0$. In such a division, every player thinks that his/her assigned piece is not smaller than the largest piece by ϵ. Section 7.2 of Brams and Taylor (1996) provides an idea for designing a procedure for ϵ-envy-free cake division, using a moving-knife. In Section 2, we give a procedure based on this idea. However, we will find that the idea does not apply to chore division. Therefore in Section 3 we provide a new moving-knife procedure for ϵ-envy-free cake division, whose idea can be easily applied to chore division. Zwicker (1997) gives the notion of "multi-fair division", which unifies both cake division and chore division. Our new procedure is further generalized to obtain an ϵ-multi-fair division. In Section 4, we define an "envy-free division with unequal ratios". We show that the given procedures can be used for approximately obtaining such a division. Finally, Section 5 concludes this paper.

As notations, we denote by $\mathcal{C}$ the whole set to be divided. Player i uses measure m_i on $\mathcal{C}$. All m_i are supposed to be additive, non-atomic probability measure defined on some common σ-algebra of subsets of $\mathcal{C}$. Without loss of generality, we let $n \geq 2$ and $\epsilon \in (0, 1)$.

2. A procedure of Brams and Taylor

Brams and Taylor (1996) suggest the following moving-knife idea to get an approximate division. It allows a player to reenter the division process, i.e., a player can call cut again and again. To put it concretely, the process asks each player initially to call cut whenever he/she thinks such a cut yields a piece of size $1/n$, and thereafter to call cut whenever he/she thinks the new piece is larger by ϵ than the one he/she presently holds. Ties are broken at random. The idea forms the following procedure.

Let

$$K_1 = \left\lceil \frac{\ln \frac{\epsilon}{2}}{\ln \left(1 - \frac{1}{n}\right)} \right\rceil .$$

It serves as a bound for the number of iterations in the following procedure.

Procedure 1: A moving-knife procedure of Brams and Taylor

Step 0 Let $k = 1$, $A_i = \emptyset$ for $i = 1, 2, \ldots, n$ and $\mathcal{R} = \mathcal{R}' = \mathcal{C}$.

Step 1 For $i = 1, 2, \ldots, n$, let $k_i = 1$ and $A_i' = \emptyset$.

Step 2 Let a knife move from the left edge of $\mathcal{R}'$ to the right. Each player i with $k_i < 2K_1/\epsilon$ can call cut. If player j calls cut, go to Step 3. *Strategy*: Player j calls cut if $m_j(\mathcal{R}_1) = \max\{m_j(\mathcal{R})/n, m_j(A_j') + \epsilon/(2K_1)\}$, where $\mathcal{R}_1$ is the left side cake of the knife.

Step 3 Let $k_j = k_j + 1$, cut $\mathcal{R}'$ into $\mathcal{R}_1$ and $\mathcal{R}_2$ (the right side cake of the knife). Let $\mathcal{R}' = \mathcal{R}_2 \cup A_j'$ (i.e., let player j put back his/her part), and let $A_j' = \mathcal{R}_1$ (i.e., allocate $\mathcal{R}_1$ to player j).

Step 4 If there is any player i with $k_i = 1$, go to Step 2. Otherwise, go to Step 5.

Step 5 Let $\mathcal{R} = \mathcal{R}'$, $k = k + 1$, $A_i = A_i \cup A_i'$ for $i = 1, 2, \ldots, n$. If $k \leq K_1$, go to Step 1. If $k > K_1$, (randomly) give the remained part to player 1, (i.e., let $A_1 = A_1 \cup \mathcal{R}$,) go to Step 6.

Step 6 Output a division $A_1, A_2, \ldots, A_n$, where A_i is assigned to player i. End.

Let us call the process from Step 1 to Step 5 a large iteration, the process from Step 2 to Step 4 a small iteration. This procedure consists of large iterations, and allocate the remained part, which is not large than $((n-1)/n)^{K_2} \leq \epsilon/2$ in any player's measure, randomly to a player. A large iteration contains at most $n\lceil 2K_1/\epsilon \rceil$ small iterations. Since each small iteration only contains one cut, we know that a large iteration needs at most $n\lceil 2K_1/\epsilon \rceil$ cuts. Each large iteration divides a part of $\mathcal{R}$ into $A_1', A_2', \ldots, A_n'$, which is envy-free within an error of $\epsilon/(2K_1)$. In the output, player i obtains A_i, which is the union of all A_i' in all large iterations. Hence we can give the following bound.

THEOREM 2.1. *For any given $\epsilon > 0$, Procedure 1 obtains an ϵ-envy-free division with at most $\lceil 2K_1/\epsilon \rceil nK_1 = O(n^3)$ cuts.*

This procedure can not be adapted to a chore division problem, because the whole set may not be sufficient enough to be allocated and some players envy others by more than ϵ. For example, let us consider the case illustrated in Figure 1. A knife moves from right to left until a player calls cut, and a player calls cut if the left side of the knife measures $1/3$. If none calls cut, then the remained part is allocated randomly to a remaining player. Then player 1 obtains $\mathcal{R}_1$ and either player 2 or player 3 obtains $\mathcal{R}_2$. In either case, one player obtains nothing and will be envied by other players.

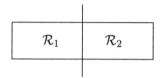

$$m_1(\mathcal{R}_1) = \tfrac{1}{3}, \quad m_2(\mathcal{R}_1) = \tfrac{2}{3}, \quad m_3(\mathcal{R}_1) = \tfrac{4}{5}$$

Figure 1: Three players' measures m_1, m_2 and m_3

In next section, a new moving-knife procedure is presented, which can be used for both cake division and chore division.

3. A new procedure for ϵ−envy-free divisions

3.1. CAKE DIVISION PROBLEM

Our new procedure lets one player cut $\mathcal{C}$ into n pieces, then let other players trim some pieces by a moving knife so that each player prefers a unique piece. We describe the procedure as follows. Let

$$K_2 = n \left\lceil \frac{\ln \tfrac{\epsilon}{2}}{\ln \left(1 - \tfrac{1}{n}\right)} \right\rceil = nK_1.$$

It serves as a bound for the number of iterations in the following Procedures 2, 3 and 4.

Procedure 2: A new moving-knife procedure

Step 0 Let $k = 1$, $A_i = \emptyset$ for $i = 1, 2, \dots, n$, $\mathcal{R} = \mathcal{C}$ and $\mathcal{R}' = \emptyset$.

Step 1 For $i = 1, 2, \dots, n$, let $k_i = 1$.

Step 2 Let player n cut $\mathcal{R}$ into n pieces A_1', A_2', $\cdots$, A_n'.
   Strategy: Cut $\mathcal{R}$ into n equal pieces.

Step 3 Let each player $i = 1, 2, \dots, n$ label one piece.
   Strategy: Each player labels a largest piece.

Step 4 If all players label different pieces (say player i labels A_i' for $i = 1, 2, \dots, n$), let $\mathcal{R}'$ be the total trimmings, go to Step 7. Otherwise go to Step 5.

Step 5 Choose a piece B labeled by at least 2 players. Let a knife move from its right edge to its left edge. Denote the left side of the knife as B_1 and the right side as B_2. Move the knife until a player i with $k_i \leq K_2/\epsilon$ calls "cut". By then go to Step 6.
Strategy: Each player i calls cut if $m_i(B_1) = (1 - \epsilon/K_2)m_i(A_i'')$, where A_i'' is a piece which i measures largest among those which player i did not label.

Step 6 Let $k_i = k_i + 1$. Cut piece B into B_1 and B_2. Let the player calling cut label another piece. Leave B_2 aside as trimmings, replace B by B_1. Go to Step 4.
Strategy: The calling cut player i moves label from B to A_i''.

Step 7 Let $k = k + 1$, $\mathcal{R} = \mathcal{R}'$, $A_i = A_i \cup A_i'$. If $k \leq K_2$, rename player n if necessary so that it represents the same player at most K_2/n times. (A_i and m_i should be renamed correspondingly.) Go to Step 1. If $k > K_2$, give $\mathcal{R}'$ (randomly) to player 1, (i.e., let $A_1 = A_1 \cup \mathcal{R}'$), go to Step 8.

Step 8 Output a division $A_1, A_2, \ldots, A_n$, where A_i is allocated to player i. End.

As notations, $\mathcal{R}$ denotes the cake part to be divided and $\mathcal{R}'$ denotes the trimmings. In Step 0, we let $\mathcal{R} = \mathcal{C}$. That is to say, the procedure starts by cutting $\mathcal{C}$. Later, the algorithm cuts trimmings by letting $\mathcal{R} = \mathcal{R}'$. Let us call the process from Step 1 to Step 7 an iteration. In each iteration, Procedure 2 divides part $\mathcal{R} - \mathcal{R}'$ of $\mathcal{C}$ into $A_1', A_2', \ldots, A_n'$. In Step 5, player i evaluates B_1 smaller than another piece A_i'', which is the largest one among those that player i did not label. This player then moves his/her label from B to A_i''. Since B is not smaller than A_i'' to player i, we know $m_i(A_i'') \leq 1/2$. Therefore $\lceil K_2/\epsilon \rceil$ cuts are sufficient for player i to be satisfied with the division $A_1', A_2', \cdots, A_n'$ within an error of $\epsilon/2K_2$. In this way, the total error in all K_2 iterations can be bounded by $\epsilon/2$. K_2 is taken large enough to ensure that the remained part $\mathcal{R}'$ after K_2/n iterations is not larger than $\epsilon/2$ to player n (the cutting player). We then rename player n so that another player will cut the cake. Finally, after K_2 iterations, the remained part is not larger than $\epsilon/2$ to all players, therefore the output division $A_1, A_2, \ldots, A_n$ is ϵ-envy-free. We leave the exact proof for this procedure to Section 3.2, where we consider a more general case.

Procedure 2 can be easily adapted to chore division problems by slightly revising Steps 3, 5 and 6 as follows.

Procedure 3: A moving-knife procedure for chore division

Step 0-2 The same as in Procedure 2.

Step 3 Let each player $i = 1, 2, \ldots, n$ label one piece.
   Strategy: Each player labels a smallest piece.

Step 4 The same as in Procedure 2.

Step 5 Choose a piece B which has no label. Let a knife move from its right edge to its left edge. Denote the left side of the knife by B_1 and the right side by B_2. Move the knife until a player i with $k_i \leq 2K_2/n\epsilon$ calls "cut". By then go to Step 6.
   Strategy: Each player i calls cut if $m_i(B_1) = (1 - \frac{n\epsilon}{2K_2})m_i(A_i'')$, where A_i'' is player i's labeled piece.

Step 6 Let $k_i = k_i + 1$. Cut piece B into B_1 and B_2. Let the player calling cut move his/her label to B_1. Leave B_2 aside as trimmings, replace B by B_1. Go to Step 4.

Step 7-8 The same as in Procedure 2.

   In each iteration, there is at least one piece only labeled by one player and therefore is not trimmed. Hence player n thinks that the size of $\mathcal{R}$ decreases at least at rate $(n-1)/n$. Therefore after K_2 iterations, each player should be approximately satisfied with the division. The proof for this procedure is also left to Section 3.2.

   In what follows, we consider a more general class of division problems, multi-fair division.

3.2. MULTI-FAIR DIVISION PROBLEM

To unify cake division and chore division, Zwicker (1997) proposes the following notion of multi-fair division. Given a set $\mathcal{C}$, integers n and l ($1 \leq l \leq n-1$). Call n pieces $A_1, A_2, \cdots, A_n$ of $\mathcal{C}$ a *multi-fair division* if each $x \in \mathcal{C}$ appears in l of the n pieces, and by assigning piece A_i to player i we have

$$m_i(A_i) \geq m_i(A_j), \quad \text{for all } i, j = 1, 2, \ldots, n. \tag{1}$$

Note that a multi-fair division does not actually form an ordinary partition except $l = 1$. Furthermore, when $l = 1$ a multi-fair division degenerates into an ordinary envy-free cake division, in which each player gets the piece he/she was assigned. On the other hand, when $l = n-1$ this problem turns to be an ordinary envy-free chore division, in which each player obtains the complement of the piece assigned to the player[1].

---

[1] For $i = 1, 2, \ldots, n$, let $\overline{A_i} = \mathcal{C} - A_i$. Then $A_i \cup \overline{A_i} = \mathcal{C}$. Therefore each $x \in \mathcal{C}$ appears in n pieces of $\{A_i, \overline{A_i}|i = 1, 2, \ldots, n\}$. Since each $x \in \mathcal{C}$ appears in $n-1$ pieces of $\{A_i|i = 1, 2, \ldots, n\}$, we know $\{\overline{A_i}|i = 1, 2, \ldots, n\}$ form an ordinary division of $\mathcal{C}$. On the other hand, from (1), we know $m_i(\overline{A_i}) = 1 - m_i(A_i) \leq 1 - m_i(A_j) = m_i(\overline{A_j})$.

Similarly, we say the pieces A_i $(i = 1, 2, \ldots, n)$ form an ϵ-multi-fair division if

$$m_i(A_i) \geq m_i(A_j) - \epsilon, \quad \text{for all } i, j = 1, 2, \ldots, n.$$

To obtain an ϵ-multi-fair division, we have the following revision of Procedure 2.

Procedure 4: A moving-knife procedure for multi-fair division

Step 0 Let $k = 1$, $A_i = \emptyset$ for $i = 1, 2, \ldots, n$, let $\mathcal{R} = \mathcal{C}$.

Step 1 For $i = 1, 2, \ldots, n$, $j = 1, 2, \ldots, l$, let $k_{i_j} = 1$.

Step 2 Let player n cut $\mathcal{R}$ into n pieces $A_1', A_2', \ldots, A_n'$.
Strategy: Cut $\mathcal{R}$ into n equal pieces.

Step 3 Let each player $i = 1, 2, \ldots, n$ label $i_1, i_2, \cdots, i_l$ on l different pieces.
Strategy: Each player labels l largest pieces.

Step 4 If each piece has l labels from l players, let $\mathcal{R}'$ be the total trimmings, go to Step 7. Otherwise go to Step 5.

Step 5 Choose a piece $\mathcal{B}$ which is labeled by at least $l + 1$ players. Let a knife move from its right edge to its left edge. Let the left side of the knife be $\mathcal{B}_1$ and the right side $\mathcal{B}_2$. Move the knife until a player s who is now labeling s_j on piece $\mathcal{B}$ with $k_{s_j} \leq \frac{2K_2}{(l+1)\epsilon}$ calls "cut".
Strategy: Each player s calls cut if $m_s(\mathcal{B}_1) = (1 - (l+1)\epsilon/(2K_2))m_s(A_s'')$, where A_s'' is a piece which s measures largest among those which s did not label.

Step 6 Let $k_{s_j} = k_{s_j} + 1$. Cut piece $\mathcal{B}$ into $\mathcal{B}_1$ and $\mathcal{B}_2$. Let the player s who called cut move label s_j from $\mathcal{B}$ to a new piece, which has no label of player s. Leave $\mathcal{B}_2$ aside as trimmings, replace $\mathcal{B}$ by $\mathcal{B}_1$. Go to Step 4.
Strategy: Player s now labels s_j on A_s''.

Step 7 Let
$$k = k + 1, \mathcal{R} = \mathcal{R}', A_i = A_i \cup \left(\cup_j \{ A_j' \mid \text{player } i \text{ has a label on } A_j' \} \right).$$
If $k \leq K_2$, rename player n if necessary so that it represents the same player at most K_2/n times. (A_i and m_i should be renamed correspondingly.) Go to Step 1. If $k > K_2$, give $\mathcal{R}'$ (randomly) to player 1, (i.e., let $A_1 = A_1 \cup \mathcal{R}'$,) go to Step 8.

Step 8 Output pieces $A_1, A_2, \ldots, A_n$, where A_i is assigned to the l players who labeled it. End.

Let us call the process from Step 1 to Step 7 a large iteration. In the same way as Procedure 2, in each large iteration, this procedure divides part $\mathcal{R} - \mathcal{R}'$ of $\mathcal{C}$ into $A'_1, A'_2, \ldots, A'_n$. There may be many small iterations from Step 4 to Step 6 inside a large iteration. We first prove the following lemmas needed later.

LEMMA 3.1. *In Step 5, $m_s(A''_s) \leq m_s(\mathcal{R})/(l+1) \leq 1/(l+1)$ for all s.*

Proof. According to the suggested strategy in Step 3, each player labels l largest pieces. Hence all l pieces labeled by s are not smaller than A''_s in s's measure. The conclusion holds because $A'_1, A'_2, \ldots, A'_n$ is a division of $\mathcal{R}$ and $m_s(\mathcal{R}) \leq m_s(\mathcal{C}) = 1$.   ∎

LEMMA 3.2. *In any large iteration of Procedure 4, there is at least one piece not trimmed in all small iterations.*

Proof. Each player can call cut only finite times, therefore there are only finite small iterations in each large iterations. Let $\mathcal{A}$ be the piece labeled by a player in the last small iteration of a large iteration. Then in the second-last small iteration, $\mathcal{A}$ is labeled only by $l-1$ players, according to the rule of the procedure. Since each trimmed piece is labeled by at least l players, we know $\mathcal{A}$ is not trimmed before the last small iteration. It is also not trimmed in the last small iteration, therefore it is not trimmed in any small iteration of the large iteration.   ∎

THEOREM 3.1. *For any given $\epsilon > 0$, Procedure 4 obtains an ϵ-envy-free multi-fair division with at most $\lceil 2K_2/(\epsilon(l+1)) \rceil n \min\{l, n-l\} K_2$ cuts.*

Proof. In each large iteration, the following relations hold:

$$m_s(A'_i) \geq m_s(A''_s) - \frac{(l+1)\epsilon}{2K_2} m_s(A''_s)$$

$$\geq m_s(A'_j) - \frac{(l+1)\epsilon}{2K_2} m_s(A''_s) \geq m_s(A'_j) - \frac{\epsilon}{2K_2}$$

for all i, j and player s who gives one label on A'_i but no label on A'_j. The first inequality is because of the suggested strategy in Step 5; the second inequality is because A''_s is the largest among all these pieces that player s did not label and the third inequality uses Lemma 3.1. Therefore after K_2 large iterations, finally $m_s(A_s) \geq m_s(A_j) - \epsilon/2$, where A_s and A_j are respectively the assigned pieces to player s and j by the procedure. By Lemma 3.2, we know that, in player n's measure, $\mathcal{R}$ decreases at least at rate $(n-1)/n$. Furthermore, according to the definition of K_2, we have

$$\left(1 - \frac{1}{n}\right)^{\frac{K_2}{n}} \leq \frac{\epsilon}{2}.$$

Since we rename player n in Step 7, we affirm that after K_2 large iterations, each player thinks $\mathcal{R}$ is not larger than $\epsilon/2$. Therefore the result is ϵ-multi-fair even we give $\mathcal{R}$ in the final large iteration to player 1.

In each large iteration, each player i calls cut at most $\lceil 2K_2/(\epsilon(l+1)) \rceil$ times for each of i's labeled piece, hence the number of cuts is at most $nl\lceil 2K_2/(\epsilon(l+1)) \rceil$. Therefore the total cut number can be bounded by $\lceil 2K_2/(\epsilon(l+1)) \rceil nlK_2$. In Step 6, player s moves a label from one piece to another piece. Hence we can also analyze the number of cuts by each unlabeled piece of each player in a similar way. In this way, we can give another bound $\lceil 2K_2/(\epsilon(l+1)) \rceil n(n-l)K_2$, which shows the conclusion. ∎

It is not difficult to check that when $l = 1$ Procedure 4 works in the same way as Procedure 2, and when $l = n - 1$ Procedure 4 works in the same way as Procedure 3. Hence we have the following bounds.

THEOREM 3.2. *The number of necessary cuts in Procedure 2 is bounded by* $\lceil K_2/\epsilon \rceil nK_2 = O(n^5)$; *The number of necessary cuts in Procedure 3 is bounded by* $\lceil 2K_2/\epsilon \rceil K_2 = O(n^4)$.

Comparing Theorem 3.2 with Theorem 2.1, we know that Procedure 2 is worse than Procedure 1 in cake divisions. However, Procedure 3 is very good in solving ϵ-envy-free chore divisions.

4. Approximate envy-Free division in unequal ratios

Ordinarily envy-freeness refers to a problem that all players divide the cake in the same ratio $(1/n)$. More generally, we now consider an envy-free division in unequal ratios. A sequence of positive numbers $\langle \alpha_1, \alpha_2, \dots, \alpha_n \rangle$ satisfying $\alpha_1 + \alpha_2 + \cdots + \alpha_n = 1$ is called an *entitlement sequence*. An envy-free division $A_1, A_2, \dots, A_n$ with entitlement sequence $\langle \alpha_1, \alpha_2, \dots, \alpha_n \rangle$ is a division in which player i receives A_i and

$$\frac{m_i(A_i)}{\alpha_i} \geq \frac{m_i(A_j)}{\alpha_j}, \quad \text{for } i, j = 1, 2, \dots, n, \tag{2}$$

where m_i is the measure on the cake of player i. A fair division with an entitlement sequence (Barbanel (1995), Shishido and Zeng (1999)) is a similar concept, which requires $m_i(A_i) \geq \alpha_i$. It is easy to show that an envy-free division with an entitlement sequence is also a fair division with the same entitlement sequence, but the contrary part does not hold.

Given $\epsilon > 0$, a division $A_1, A_2, \dots, A_n$ is called ϵ-envy-free with entitlement sequence $\langle \alpha_1, \alpha_2, \dots, \alpha_n \rangle$ if

$$\frac{m_i(A_i)}{\alpha_i} \geq \frac{m_i(A_j)}{\alpha_j} - \epsilon, \quad \text{for } i, j = 1, 2, \dots, n.$$

Section 7.6 of Brams and Taylor (1996) considers a special case[2] of this problem when each α_i is a rational number. The authors argue that all the procedures for ordinary envy-free divisions can be directly applied to the case, but no bound for the number of cuts is given there.

Now we apply Procedure 1 to obtain an approximate envy-free division for a general situation where some α_i may be irrational numbers. A bound will be given for the number of cuts in the procedure. It consists of two stages. First, we use suitable rational numbers to approximate the entitlement sequence $\langle \alpha_1, \alpha_2, \ldots, \alpha_n \rangle$. This can be done as follows. Let $\alpha_{\max} = \max_i \alpha_i$ and $\alpha_{\min} = \min_i \alpha_i$. Then $0 < \alpha_{\min} \leq \alpha_{\max} < 1$. Let

$$N = \lceil \frac{\alpha_{\max}}{\alpha_{\min}^3} \frac{5}{\min\{\epsilon, 1\}} \rceil, \quad \text{and} \quad N_i' = \lfloor \alpha_i N \rfloor \quad \text{for} \quad i = 1, 2, \ldots, n.$$

Then $N - n < \sum_{i=1}^n N_i' \leq N$. Therefore $k = N - \sum_{i=1}^n N_i'$ is an integer in $[0, n)$. Let

$$N_i = \begin{cases} N_i', & \text{for } i = 1, \ldots, n-k, \\ N_i' + 1, & \text{for } i = n-k+1, \ldots, n. \end{cases}$$

We shall use $\langle \frac{N_1}{N}, \frac{N_2}{N}, \ldots, \frac{N_n}{N} \rangle$ to approximate $\langle \alpha_1, \alpha_2, \ldots, \alpha_n \rangle$.

LEMMA 4.1. *It holds that* $|N_j/N_i - \alpha_j/\alpha_i| \leq \epsilon \alpha_{\min}/2, \forall i, j = 1, 2, \ldots, n.$

Proof. By the definition of N_i, we have

$$\left| \frac{N_i}{N} - \alpha_i \right| \leq \left| \frac{N_i - N_i'}{N} \right| + \left| \frac{N_i' - \alpha_i N}{N} \right| < \frac{1}{N} + \frac{1}{N}. \qquad (3)$$

Furthermore, from

$$\begin{aligned} N_j \alpha_i - N_i \alpha_j &\leq (N_j' + 1)\alpha_i - N_i' \alpha_j = \alpha_i + \alpha_i N_j' - \alpha_j N_i' \\ &\leq \alpha_i + \alpha_i(\alpha_j N) - \alpha_j(\alpha_i N - 1) = \alpha_i + \alpha_j, \\ N_j \alpha_i - N_i \alpha_j &\geq N_j' \alpha_i - (N_i' + 1)\alpha_j = N_j' \alpha_i - N_i' \alpha_j - \alpha_j \\ &\geq (\alpha_j N - 1)\alpha_i - (\alpha_i N)\alpha_j - \alpha_j = -(\alpha_i + \alpha_j), \end{aligned}$$

we know $|N_j \alpha_i - N_i \alpha_j| \leq \alpha_i + \alpha_j \leq 2\alpha_{\max}$. On the other hand, since $n \geq 2$, we know $\alpha_{\min} \leq 1/2$. From the definition of N, we know $N \geq 10/\alpha_{\min}$, hence $\alpha_i(\alpha_i - 2/N) \geq 4\alpha_i^2/5$. Therefore,

$$\left| \frac{N_j}{N_i} - \frac{\alpha_j}{\alpha_i} \right| = \frac{\frac{|N_j \alpha_i - N_i \alpha_j|}{N}}{\alpha_i \frac{N_i}{N}} \leq \frac{\frac{2\alpha_{\max}}{N}}{\alpha_i(\alpha_i - \frac{2}{N})}$$

$$\leq \frac{\frac{2\alpha_{\max}}{N}}{\frac{4\alpha_i^2}{5}} \leq \frac{5\alpha_{\max}}{2N\alpha_{\min}^2} \leq \frac{\epsilon}{2}\alpha_{\min},$$

---

[2] Our definition is consistent with their definition, because there exists a partition of any A_i into n pieces such that all the pieces are of the same size to all the n players (Neyman (1946)).

where the first inequality is from (3) and the last inequality is from the definition of N.   ∎

Imagine $N_i - 1$ clones of player i for $i = 1, 2, \ldots, n$. The second stage is to apply Procedure 1 to let total $N_1 + N_2 + \cdots + N_n = N$ players divide the cake so that the result is approximately envy-free within an error of $\epsilon' = \epsilon \alpha_{\min}/(2N)$. Finally, player i obtains all the parts of his/her clones in addition with his/her own part for $i = 1, 2, \ldots, n$. Denote the total part of player i by A_i. Then we have:

LEMMA 4.2. *Division* $A_1, A_2, \ldots, A_n$ *forms an* ϵ-*envy-free with entitlement sequence* $\langle \alpha_1, \alpha_2, \ldots, \alpha_n \rangle$.

Proof. After applying Procedure 1, suppose that in player i's measure, the piece which is the smallest among those obtained by player i and his/her clones is A_i', and the piece which is the largest among those obtained by player j and his/her clones is A_j', then $m_i(A_i') \geq m_i(A_j') - \epsilon \alpha_{\min}/(2N)$. Hence

$$\frac{m_i(A_i)}{N_i} \geq m_i(A_i') \geq m_i(A_j') - \frac{\epsilon \alpha_{\min}}{2N} \geq \frac{m_i(A_j)}{N_j} - \frac{\epsilon \alpha_{\min}}{2N}$$

Therefore

$$
\begin{aligned}
m_i(A_j) &\leq N_j \left(\frac{m_i(A_i)}{N_i} + \frac{\epsilon \alpha_{\min}}{2N} \right) \\
&= \frac{N_j}{N_i} m_i(A_i) + \frac{\epsilon}{2} \frac{N_j}{N} \alpha_{\min} \leq \frac{\alpha_j}{\alpha_i} m_i(A_i) + \frac{\epsilon \alpha_{\min}}{2N} + \frac{\epsilon \alpha_{\min}}{2N} \\
&\leq \frac{\alpha_j}{\alpha_i} m_i(A_i) + \epsilon \alpha_j = \alpha_j \left(\frac{m_i(A_i)}{\alpha_i} + \epsilon \right),
\end{aligned}
$$

where the second inequality is from Lemma 4.1. The above relations show that $m_i(A_i)/\alpha_i \geq m_i(A_j)/\alpha_i - \epsilon$.   ∎

THEOREM 4.1. *Given* $\epsilon > 0$ *and entitlement sequence* $\langle \alpha_1, \alpha_2, \cdots, \alpha_n \rangle$. *Let*

$$\alpha_{\min} = \min_i \alpha_i, \quad \alpha_{\max} = \max_i \alpha_i,$$

$$N = \left\lceil \frac{\alpha_{\max}}{\alpha_{\min}^3} \frac{5}{\epsilon} \right\rceil, \quad \epsilon' = \frac{\epsilon}{2N} \alpha_{\min}, \quad K_3 = \left\lceil \frac{\ln \frac{\epsilon'}{2}}{\ln \left(1 - \frac{1}{N} \right)} \right\rceil.$$

There is a procedure to obtain an ϵ-*envy-free division with the given entitlement sequence, in which the number of necessary cuts is bounded by* $\lceil 2K_3/\epsilon' \rceil N K_3$.

Proof. The conclusion follows from Theorem 2.2 and Lemma 4.1.   ∎

The above argument can be similarly applied to chore divisions. Since Procedure 1 is not suitable for chore divisions, we have to apply Procedure 3 to replace Procedure 1. Let

$$K_4 = N \left\lceil \frac{\ln \frac{\epsilon'}{2}}{\ln \left(1 - \frac{1}{N}\right)} \right\rceil,$$

where N and ϵ' are defined in Theorem 4.1. By Theorem 3.2, the number of necessary cuts for this kind of problems can be bounded by: $\lceil 2K_4/\epsilon' \rceil K_4$.

5. Concluding Remarks

Since procedures for exactly fair divisions need too many cuts, finding practical approximately fair procedures is important. Cake division and chore division problems are similar but different in some ways. Although Brams and Taylor's approximate procedure is good for cake divisions, it does not work for chore divisions. Hence this paper provides a new ϵ-envy-free procedure, which can be used to obtain both cake division and chore division. The procedure is further generalized to a multi-fair division, which unifies cake division and chore division. Finally, the studied procedures are applied to obtain ϵ-envy-free division in unequal ratios, where a bound is given for the number of cuts.

Recently, Su (1999) gives another procedure for approximately fair division, which works well for both cake division and chore division. His procedure only needs the minimal number of cuts $(N - 1)$, but it involes many labelling steps.

There are some other ways to consider approximately fair division problems. For example, Robertson and Webb (1995) and (1998) consider the following two approximating problems: how much of the cake can be guaranteed to all the players if a restricted number of cuts is used? How many players can be guaranteed $1/n$ of the cake if a restricted number of cuts is used?

Acknowledgements

The author is very grateful to J. B. Barbanel, S. J. Brams, H. R. Mihara, R. R. Ranade, F. E. Su, A. D. Taylor and W. S. Zwicker for their kindness to discuss many problems.

References

Barbanel, J. B. (1995) Game-Theoretic Algorithms for Fair and Strongly Fair Cake Division with Entitlements, *Colloquium Mathematicum*, **Vol. 69**, pp. 59-73.

Brams, S. J. and Taylor, A. D. (1995) An Envy-Free Cake Division Protocol, *American Mathematical Monthly*, **Vol. 102**, pp. 9-18.

Brams, S. J. and Taylor, A. D. (1996) *Fair Division: From Cake-Cutting to Dispute Resolution*, Cambridge University Press. f

Brams, S. J., Taylor, A. D. and Zwicker W. S. (1997) A Moving-Knife Solution to the Four-Person Envy-Free Cake Division Problem, *Proceedings of the American Mathematical Society* **Vol. 125**, pp. 547-554.

Neyman, J. (1946) Un Theoreme D'Existence, *C. R. Acad.Sci. Paris* **Vol. 222**, pp. 843-845.

Peterson E. and Su F. E. (1988) Exact Procedures for Envy-Free Chore Division, *preprint*.

Robertson J. M. and Webb W. A. (1995) Approximating Fair Division with a Limited Number of Cuts, *Journal of Combinatorial Theory, Series A*, **Vol. 72**, pp. 340-344.

Robertson J. M. and Webb W. A. (1998) *Cake-Cutting Algorithms: Be Fair If You Can*, A. K. Peters.

Shishido H. and Zeng D.-Z. (1999) Mark-Choose-Cut Algorithms for Fair and Stronly Fair Division, *Group Decision and Negotiation*, **Vol. 8**, No. 2, 125-137.

Steinhaus H. (1948) The Problem of Fair Division, *Econometrica*, **Vol. 16**, pp. 101-104.

Su F. E. (1999) Rental harmony: Sperner's lemma in fair division, to appear in *American Mathematical Monthly*.

Zwicker W. S. (1997) Private Communication.

Index

THEORY AND DECISION LIBRARY

SERIES C: GAME THEORY, MATHEMATICAL PROGRAMMING AND OPERATIONS RESEARCH
Editor: S.H. Tijs, *University of Tilburg, The Netherlands*

18. T. Parthasarathy, B. Dutta, J.A.M. Potters, T.E.S. Raghavan, D. Ray and A. Sen (eds.): *Game Theoretical Applications to Economics and Operations Research.* 1997   ISBN 0-7923-4712-9
19. A.M.A. Van Deemen: *Coalition Formation and Social Choice.* 1997
                                                              ISBN 0-7923-4750-1
20. M.O.L. Bacharach, L.-A. Gérard-Varet, P. Mongin and H.S. Shin (eds.): *Epistemic Logic and the Theory of Games and Decisions.* 1997
                                                              ISBN 0-7923-4804-4
21. Z. Yang (eds.): *Computing Equilibria and Fixed Points.* 1999
                                                              ISBN 0-7923-8395-8
22. G. Owen: *Discrete Mathematics and Game Theory.* 1999
                                                              ISBN 0-7923-8511-X
23. I. Garcia-Jurado, F. Patrone and S. Tijs (eds.): *Game Practice.* 1999
                                                              ISBN 0-7923-8661-2

KLUWER ACADEMIC PUBLISHERS – DORDRECHT / BOSTON / LONDON